Patrick's Florilegium

Patrick Harding

Patrick's Florilegium

Text and photographs
Patrick Harding

With botanical illustrations by
Jean Binney

The Hallamshire Press
2005

Patrick's other books include:

How to Identify Edible Mushrooms (1996)
with Tony Lyon and Gill Tomblin, 8th Printing, HarperCollins
Collins Gem Mushrooms (1996, 2003 2nd ed.) HarperCollins
How to Identify Trees (1998) 4th Printing. HarperCollins
Wild Flowers of the Peak District (2000) Hallamshire Press
The Magic of Christmas (2002, 2004 3rd ed.) Metro (Blake Publishing)

In preparation for publication in 2006:
Collins Need to Know? Mushroom Hunting HarperCollins

*Every effort has been made to obtain permission from the relevant publishers
to reproduce the poems in this book. All authors and publishers, where known,
are cited in the bibliography on pages 139–142.*

© Patrick Harding 2005

Published by The Hallamshire Press for Patrick Harding

The Hallamshire Press is an imprint of
Hallamshire Publications Limited
Porthmadog, Gwynedd

Designed and Typeset by Hallamshire Publications Limited
Printed in Spain by Edelvives, Zaragoza

British Library Cataloguing in Publication Data:
   A catalogue record for this book is available from the British Library

ISBN: 1874718 660

Frontispiece: Emerging sycamore leaves

Contents

Acknowledgements   6
Introduction   7

The Species   13

Bibliography   139
Index   142

Acknowledgements

The information for this book has come from a wide range of sources. In some cases I have received assistance from my adult students who have pointed me towards articles, books or web sites. Please accept my thanks for your help. My wife Jean and teenagers Martin and Bryony have had to put up with me as I raced to keep my deadlines; in addition a big thank you to Jean for the wonderful illustrations portrayed in the book. I am also in debt to my niece Charlotte Pover who checked the proofs in her usual friendly and efficient manner, and for the help and advice from my publisher, Pauline Climpson along with Jenny Sayles and Pat Whitehead.

Introduction

This book has had a very long gestation period. Fifty years ago I came back to this country as a nine-year-old from Egypt, where I had spent most of my formative years while my mother taught at an army school and my father forecast the weather for the RAF. The Suez Crisis was brewing and my years of playing in the sand and swimming in the Bitter Lakes had come to an end. Suddenly there were no more scorpions, no burning sunshine; but a very green and pleasant land. The first mistake I made was to walk on a cow pat; my previous encounters with the dung of donkeys and camels had been met with a more stern resistance. Half a century on I have not quite lost the knack of putting my foot in it.

Our new home at RAF Wyton was in the county of Huntingdonshire, since done away with by meddling politicians, and my primary school was at Houghton, a mile away on a local bus. When the weather was good I saved the bus fare and walked back. On my journeys I began to appreciate the changing seasons and to learn about the flowers that grew in the roadside verges. My mother had an excellent knowledge of plant names and I was eager to learn. I remember bringing a vase of wild flowers to school one day, each with a name-tag hanging from its stem. I was given a gold star by my teacher, a tough northerner who was strict but fair: a great role model. He surprised us all, and no doubt the headmaster, some months later when he distributed a chocolate bar to everyone in the class to celebrate the news that his first article had just been accepted for publication. His debut piece was duly published in *Weekend*, the nearest the 1950s had to compare to our current crop of lads' magazines.

The country boarding school where I received my secondary education was keen to ensure that its pupils stayed within bounds; the castle that formed the main building had restrained Catherine of Aragon when Henry had grown tired of her. Unlike my contemporaries I had a wonderful excuse to roam far and wide—I was recording the local wild flowers for a new Flora of Huntingdonshire. Among the rarities that I turned up on my Sunday rambles was the crested cow wheat, a plant I have not seen since! During my teens my parents moved to Bracknell where, in the holidays, I continued to seek out unusual flowers, this time for the Flora of Berkshire. In an area of mostly clay or sandy soils I was delighted to find hoary plantain, a plant restricted to calcium-rich sites. It was growing in a local churchyard and I realised that it was possible that the plant's calcium supply came from a macabre source.

In 1964 I scraped enough A-level passes to get into university and study botany. In those days applying for university entry was a rather more laid-back affair than it is today. I put my choices down in alphabetical order and went to Bangor as I could not spell Aberystwyth. The first thing I did with my grant money was to buy a single-lens reflex camera, so that I could take close-up photographs of flowers. The camera was a Praktika, made in East Germany: cheap, heavy and built like a tank—I dropped it from a tree but it continued to work. So I began to amass a

Viola sp.

collection of over 20,000 slides. Little did I know then just how useful they would become as illustrations for my lectures and books.

At the end of my first year at Bangor my botany mark was a lowly 40% (I never could cut thin sections and as for trying to draw what I couldn't see…), but I had by then discovered the delights of agricultural botany. Trying to grow plants on old copper mines was much more interesting than anatomy. After Bangor I moved up (and right a bit) to York to study for my doctorate. The local paper printed a caption under my photograph that read 'Patrick Harding testing plants for coal tips'. There I was at a new university in the 1960s and all I was doing with grass was using it to cover up the mounds of spoil waste from the collieries of South Yorkshire. Arthur Scargill is alleged to have declared the work a waste of money as he didn't see anything wrong with black-coloured hills, but I carried on anyway. One day my professor spoke to me about a local natural history evening class and gave me a phone number. I felt that I had moved on from such basic learning, but phoned to keep the boss happy. I realised too late that the class was not short of pupils but needed a teacher.

The class met in the nearby town of Selby and not having learnt to drive at the time I had to travel there by bus. As the group had already done plants I had to start with British animals. On the way to the bus one evening I picked a moribund hedgehog off the road, sensing a good visual aid for the class. As I put the prickly specimen down on my case so that I could pay for my bus ticket, half a dozen fleas headed rapidly for the exit: the driver said nothing by mouth but his eyes said it all. A few weeks later I took my class through all the common British butterflies and then paused for questions. 'What's the purpose of a butterfly?' was the sole query. In 1969 I had yet to learn of the importance of butterflies in the theory of chaos, so I gave the more mundane reply that not everything had to have a purpose.

The Selby class was organised by the Workers' Educational Association and I still teach for the WEA in both Sheffield and Chesterfield. Along with most people who work at the chalkface, despite ordinary chalk no longer being allowed for health reasons, I am sceptical of the huge increase in administration and resultant non-teaching staff that forms part of my workload. I was recently requested to fill in a form where two pages were reserved for the intimate details of my training as a teacher. Interestingly, there was no place to put 'I learnt on the job'. I also learnt from my mother, who once recounted that on her class inspection as part of her teaching diploma one of the children had asked whether she could tell him the scientific name of the plant on her desk. Realising that no-one, including herself and the inspector, had any idea of the answer, she made one up and promptly passed with flying colours.

Armed with my doctorate from York I took off across the water to another new university, this one at Coleraine in Northern Ireland, just as the 'Troubles' were starting. Here I gained valuable experience teaching plant ecology and evolutionary genetics to undergraduates. Two important developments took place during my seven years in Ulster. The first was that I started taking photographs of fungi and, after finding a giant puffball that fed lots of friends, I began my love affair with edible mushrooms. (I also had lots of other love affairs, but these are outside the scope of this book.) The second development was that I started giving illustrated talks to local societies. Slowly I realised that for me there was more to further education than lecturing to 18-year-olds. On the day that I left Coleraine two arts students in the lift asked whether I was the lecturer who believed in evolution. Perhaps my mission to Ulster remains unfulfilled.

My move to the extramural department at Sheffield University was to give me a freedom rarely found in any other area of education today. As I built up one of the largest natural science-based adult programmes in the country I was able to develop new interests and link them to my teaching. The result of this was that I began to teach courses on the identification of fungi and explored the reasons behind plant names in courses on herbal medicine. Partly through my interest in fungi I began to learn more about trees. I also came to realise that garden plants could be interesting especially when one examined the life and times of the plant hunters who had been responsible for their introduction. In addition to my Sheffield work I started to take residential courses in centres all round Britain. One of my weekends on trees was as likely to include a session on poetry and prose as it was one on tree biology. Oh the beauty of not being hidebound by a curriculum!

On a fungus weekend that I tutored in Cumbria one of my 'students' turned out to be Barry Fox, technology editor for *New Scientist*. Barry wrote a cracking piece based on the weekend for his periodical and I supplied the photographs. Three weeks later the natural

Magic mushroom

9

history editor phoned from Harper Collins and the idea of the edible fungus book was hatched. Had my old English teacher realised that I would go into print he would have fallen from his bike as in the 1960s he was sufficiently pleased to have pushed me through O-level. Thanks to the help from my great friend and co-author Tony Lyon plus the superb artwork by Gill Tomblin, *How to Identify Edible Mushrooms* has remained a steady seller for nine years. The same year, 1996, also saw the publication of my *Gem Guide to Mushrooms and Toadstools*, which by 2004 (in its new edition *Gem Mushrooms*) was still selling over 10,000 copies a year. In 1998 *How to Identify Trees* joined its mushroom stablemate on the bookshelves.

In 1986 I married Jean Binney, one of my adult students—another of the joys of adult education is that one can have social and other intercourse with the pupils. Wife begins at forty, I had been told, and so it turned out! Jean enrolled on a number of University courses, including those taught by a local tutor, Valerie Oxley. Valerie taught botanical illustration and the success of her classes led to the university setting up a diploma in the subject, with Valerie teaching the art while I taught the botany. Jean was among the first group of students to be awarded the diploma. Out of the wonderful watercolour work produced by Valerie's students came the idea of a book about local wild flowers. In 2000 *Wild Flowers of the Peak District* was published by the Hallamshire Press. My job was simply to fill the gaps between the artwork, but it gave me the chance to write something about the medicinal and other uses of the illustrated plants.

By this time I had taken very early retirement from the university and had launched into the world of being a freelance lecturer, author and media person. I had also found a new interest: Christmas. This started while I was still at the university. A group of staff at the department decided that it would be nice to develop a drab area above the boiler room into a roof garden. The university agreed to match funding, so we had to come up with money-making ideas. My contribution was a special lecture on the natural history of Christmas. I knew enough about holly, ivy, robins and Christmas trees, but did not anticipate the interest: I had to give the talk twice to satisfy the demand for tickets. As the talk became a favourite for local societies so I learnt more about all the other aspects of Christmas.

Green man carved on a tree in Ireland

Roger Highfield, science correspondent of the *Daily Telegraph*, heard about this strange Father Christmas-clad lecturer who gave talks about the science behind Christmas. I was interviewed in November and during the photo shoot was locked out of my house. I got some funny looks as I walked down the road in full gear to pick up a spare key from my parents-in-law. The article led to lots of media interest and finally to my fifth book. To date this has gone through three editions, all with a different name. Its latest incarnation is *The Magic of Christmas* (2004) published by Metro.

One of the beauties of my work is that it follows the seasons. During May, June and July I teach field-based courses on flower identification, herbal medicine and trees. August is reserved for family holidays. September and October are given over to courses on fungi. During November and December life is preoccupied with talks, courses and media events on the subject of Christmas. This leaves January, February, March and April for my writing. What could follow mushrooms, trees, Peak District flowers and Christmas? The answer is this book. For some 30 years I have collected information about many of the species that I teach people to recognise.

Names interest me. In my youth it was the twitcher's thrill of adding a new name to the tick-list, but in my dotage I have discovered the pleasures of etymology. Good jokes in Christmas crackers are as rare as honest politicians, but here is one: Question—'What is the difference between an etymologist and an entomologist?' Answer—'The first one can tell you'. In this book I have had the luxury of being able to look in depth at the origin of both Latin and common names. The book also investigates the uses of plants. These are not restricted to matters of health, although many of the species covered in this book have not just a historic but a present and probable future medicinal use.

Education can be as dry as Martini, though just as intoxicating. I have tried in this book to mix facts with anecdotes and humour. This is not a case of 'dumbing down' but of the realisation (I wonder if it is taught in teacher training) that humour, like enthusiasm, keeps the reader or audience awake and provides a peg on which to hang the knowledge that goes with the joke. Dry and academic do not have to go hand in hand.

Cape gooseberry

Working with my wife Jean, who has produced the wonderful watercolour illustrations, has been another pleasure during the book's production. Photographs may be pretty, but good botanical illustration carries so much more accurate, detailed information.

The book covers a wide range of species; not all are plants and not all are native. The inclusion of garden plants, introductions and lower plants, including a moss, seaweed and fern, together with fungi and a lichen, has been a deliberate attempt to cross boundaries. The only drawback to the subject matter has been in deciding on the book's title. The book is not simply another identification guide. One working title was *Flowers, Fruits, Fungi, Ferns and Fucus*, but alliteration can go too far and the F-word is much overused. *Patrick's Potpourri* is perhaps too literary and reminiscent of my former Ulster MP's cry, 'No popery here'. So it is *Patrick's Florilegium*, the latter word meaning an anthology and like the book, not being restricted to flora. Interestingly, the word anthology, now used for collections of poetry or prose, comes from two Greek words: *anthos*, meaning a flower, and *logia*, a collection. As my interest in nature started with flower collections, the title seems particularly apt.

Boot with house leeks

12

The Species

Bladder Wrack	14
Bog Moss	18
Coltsfoot	22
Deadly Nightshade	26
Dove Tree	30
Elder	34
Ergot	38
Eyebright	42
Fly Agaric	46
Foxglove	50
Horse Chestnut	54
Ivy-leaved Toadflax	58
Juniper	62
Lesser Celandine	66
Liquorice	70
Magic Mushroom	74
Male Fern	78
Meadow Saffron	82
Melilot	86
Oak Moss	90
Oxford Ragwort	94
Pineapple Weed	98
Porcelain Fungus	102
Shaggy Ink Cap	106
Sycamore	110
Tulip Tree	114
Valerian	118
White Bryony	122
Wild Arum	126
Wormwood	130
Yew	134

Bladder Wrack
Fucus vesiculosus

Many botanists are sadly 'all at sea' when it comes to identifying more than a handful of the many hundreds of species of seaweed that inhabit Britain's coastline. Seaweeds are classified as Algae, a primitive group of plants that include single-celled species such as those that make up much of the marine phytoplankton. The multicellular structure of a seaweed is more complex but the level of differentiation into cell and tissue types is considerably below that found in the flowering plants. Seaweeds such as those in the genus *Fucus* lack a specialised transport system, do not possess separate stem, leaf and root tissues, and do not produce flowers, fruits or seeds.

Some algae inhabit terrestrial habitats such as tree trunks and paved paths, but most are aquatic. Many species live in freshwater but the greatest diversity of algae is to be found in the marine environment. Simple classification systems still separate the marine algae into three groups on the basis of their colour: green (Chlorophyceae), red (Rhodophyceae) and brown (Phaeophyceae). This may not appear to be a very scientific approach but the colour of a seaweed is strongly influenced by the proportion of photosynthetic and other pigments that it contains. These in turn largely determine the microhabitat of the species. Most green seaweeds inhabit the upper reaches of the shore where they are only covered by seawater for short periods each day. In contrast, those such as the brown *Fucus* species inhabit deeper water or rocks only briefly exposed at low tide. Many seaweeds, including the wracks, have a mucilaginous covering that serves to reduce desiccation during the time when the plants are not covered by water.

Bladder wrack grows in dense stands attached to rocky outcrops in the middle shore zone. The repeatedly forked, dark olive-brown, strap-like blades are perennial and can grow up to a metre in length. The wavy-edged blades have a thick flattened midrib. Along the blade, pea-sized vesicles (air bladders) arise at irregular intervals, usually in pairs with one bladder on each side of the midrib. These help to keep the plant apex floating on or near the water surface, and therefore nearer the sunlight, at high tide. Popping the bladders is an essential part of a seaside holiday for children of all ages, although plants of the more exposed upper shore regions may lack bladders.

The lower region of the midrib is rounded in cross-section and devoid of blade-like edges. It is known as the stipe and is attached to rocks and stones by a disc-shaped, root-like 'holdfast'. At the other end of the plant, the tips of some of the blade branches take on a swollen, sultana-like appearance. These receptacles are covered with tiny pores that are the openings of flask-like structures (conceptacles) buried within the spongy tissue. Bladder wrack plants are either male or female and the conceptacles of female plants produce masses of single-celled eggs that are released through the pores into the sea. Male conceptacles give rise to sperm cells that are also released into the sea and swim towards the egg cells with a whiplash motion of two hair-like projections. The release of the sex cells is synchronised with a period of high tides. This makes it

Close up of bladder wrack showing the paired bladders

more likely that fertilisation will occur and, as the tide cycle is influenced by the moon, it gives a new meaning to the term 'lunatic fringe'.

Closely related species include toothed wrack (*Fucus serratus)*, which normally inhabits the lower shore. It has more golden-brown, broader blades that lack bladders but have serrated edges and more flattened receptacles. Spiral wrack (*Fucus spiralis*) is typically found higher up the shore than bladder wrack. It has similar untoothed blades but these are twisted near the apex, lack bladders and bear very swollen, globular receptacles. During violent storms *Fucus* plants are torn from their substrate and cast onto the shore along with rather more substantial brown seaweeds in the genus *Laminaria*. Seaweeds marooned on the shore in this way are known as wracks. Dried wrack was formerly an important organic fertiliser to farmers and gardeners in coastal communities. A wrack gatherer was depicted on one of the first postage stamps issued by the Channel Isles after their liberation from German occupation in 1945. In parts of Scotland species of *Fucus* are eaten by sheep as a supplement to their normal fodder.

Historically the wracks had other roles to play in a range of industrial processes and medical preparations. From the late 17th century to the middle of the 19th century the ash resulting from the burning of what were collectively called kelps (brown seaweeds, mostly species of *Fucus* and *Laminaria*) provided chemicals for the production, among other commodities, of glass. At its peak 60,000 people derived a living from kelping in Scotland alone. The process involved the cutting, collection and drying of the weed followed by burning it in a kelp pit. During my regular visits to the Isles of Scilly, where kelp burning started in 1684, I enjoy seeking out examples of the remaining kelp pits—small stone-lined depressions close to the seaweed-covered shore.

The resulting ash from the burnt kelp was often termed soda ash, although sodium was only one of its constituents. The soda content varied with the seaweed species used, bladder wrack being one of the richest. The ash also contained potash and iodine, though the yield of the latter was much higher from species of *Laminaria*. Over 20 tons of wet weed had to be collected to make 1 ton of ash.

The raw materials for glass making are sand (silica), soda and lime. Silica has to be heated to a temperature in excess of 1700°C before it will melt. The addition of soda ash (sodium carbonate) acts as a flux, and helps to reduce the melting point of silica to nearer 850°C. This made the process simpler and cheaper. Medieval glass-makers also used glassworts (species of flowering plant in the genus *Salicornia* that concentrate salts from the saline mud flats where they grow). Glassworts are found on British coasts but more were imported from Spain under the name of barilla. Other sources of flux included potash (potassium carbonate) produced from burnt trees, bracken and male fern (see pages 78–81). Kelp was a ready substitute, especially during the Napoleonic Wars, when imports of barilla were cut off. During this same time the French lost not only their supplies of kelp from Scotland but also barilla from Spain and this prompted the search for an alternative source of soda ash or potash. The chemist Nicolas Leblanc came up with a process for manufacturing sodium carbonate from common salt. When this was introduced to Britain in the 1820s it killed off the demand for kelp products from the glass-makers.

Soda ash was also formerly used in the manufacture of soap, when it was dispersed in water before the addition of fat, followed by prolonged boiling. Sodium carbonate produced by the Leblanc process was a cheaper alkali than that supplied from burning kelp and resulted in a sharp drop in the price of soda ash. A minor use for soda ash was in the production of alum, used in the manufacture of paints, textiles and baking powder. In the 20th century new uses were discovered for the mucilaginous alginate chemicals found in brown seaweeds such as species of *Fucus* and *Laminaria*. These alginates are still used in the manufacture of a wide range of products, including ice-cream, soup and beer.

Brown seaweeds have also had an important role to play in medicine. My home in Sheffield is close to the Yorkshire/Derbyshire border and a long way from the sea. Before the railways resulted in the rapid transportation of seafood to inland regions of Britain, the inhabitants of Derbyshire suffered more than most from a swelling of the thyroid gland referred to locally as

*Bladder wrack
and toothed wrack
with limpets*

*Disused kelp pit on
the Isles of Scilly*

'Derbyshire neck'. Lack of iodine is a principal cause of the condition we now call goitre. In 1750 kelp 'charcoal' was introduced as a cure for goitre and in 1811 the French chemist Courtois discovered a way of extracting the iodine present in kelp ash. In addition to treating goitre, iodine was found to be an excellent antiseptic.

The knowledge that kelp could be used as a source of iodine came at a good time for the seaweed burners of western Scotland, where iodine-rich brown seaweeds were plentiful, as 1811 was only a few years before the ash was no longer required in the manufacture of glass and soap. By 1846 there were twenty manufacturers of iodine in the Glasgow area alone. Chemistry finally put paid to the kelp industry when a readily available supply of extractable iodine was discovered in Chile saltpetre, originally laid down in a now long-vanished sea. Today iodine is added to table salt, thus ensuring an adequate intake. 'Thick necks' are now something of a rarity.

Powdered fucus (as the drug is known) was used not only to counter goitre but also in the form of wound dressings, where its iodine content was probably the reason for its antiseptic properties. Cotton or lint (previously made from linen) dressings are widely used on open wounds but they stick to congealed blood and their removal may damage recently formed scar tissue. Recently medicine has found a new use for seaweeds as surgical dressings. Sorbsan is a dressing woven from a thread of calcium alginate—the alginate coming from brown seaweeds. The dressing not only aids blood clotting, but can also be left on a wound where contact with blood converts it to sodium alginate, a gel that soaks up bodily fluids and is itself slowly dissolved and absorbed into the body. The remaining dry part of the dressing can be removed without damaging any newly formed scab.

> *Ring the alarum bell—Blow, wind! Come wrack!*
> *At least we'll die with harness on our back.*

Shakespeare, *Macbeth*, Act V Scene V

Bog Moss

Sphagnum spp.

Along with two other species featured in this book, bog moss is not a flowering plant and in evolutionary terms it falls between bladder wrack (page 14) and male fern (page 78). Unlike most algae, such as bladder wrack, the bryophytes (liverworts and mosses) have made the transition from water to land, although most *Sphagnum* species are still very water dependent. Unlike seaweeds, mosses consist of separate stem and leaf tissues but they lack the more complex root and transport systems of the ferns.

Of the native plants found in Britain there are about 1,500 species of flowering plant, only about 50 species of fern but over 750 species of moss. The small size of mosses and apparent difficulty in distinguishing between species has resulted in a general ignorance of the group, even among naturalists. Some small ferns are superficially moss-like, as are a number of flowering plants such as the pearlworts (*Sagina* spp.). On many occasions I have fielded complaints from gardeners who, having applied moss-killer to their lawns, bring me the surviving 'moss', which turns out to be a flowering plant. Mosses have also been confused with lichens, as will be seen on page 90.

One result of the poor general knowledge about mosses is that relatively few of them have common names—one that does is bog moss. In a manner comparable to that of eyebright (page 42), the name bog moss is used for a whole group of related species of *Sphagnum* which vary in size, colour and habitat preference. The bog mosses have played an extremely important role in the landscape and ecology of large parts of the British Isles and their products have a wide range of uses, hence the common name. The accurate naming of many of the individual *Sphagnum* species is a job for an expert but fortunately the bog mosses differ in their basic structure from all other British mosses.

Most bog mosses form dense mats which, when wet, are soft and spongy to the touch. They come in various shades of green, gold, pink, bronze or purplish-red—the colour sometimes helping in the separation of closely related species. In dry conditions the plants are tinged with white and their texture becomes crisp and brittle. The cushion-like mats consist of numerous crowded, erect stems that range (depending on species and growing conditions) from only a few centimetres to over 30cm in height. Very young plants have small, primitive root-like rhizoids at the stem base; these are absent from older plants where the stem base dies away but remains attached to the rest of the plant. A unique feature of the bog mosses is that their erect stems bear rosette-like branches in bunches (fascicles), some of which hang down and clothe the stem while others grow outwards or even upwards. The side branches are liberally clothed in overlapping, scale-like leaves that are typically longer and more pointed than those on the main stem.

The leaves on the branches have a very unusual anatomy. They are made up of a network of elongated cells containing the green pigment chlorophyll. These cells are interspersed by groups of dead hyaline cells that have pores in their walls through which water can pass. Similar hyaline-like cells are found in the outer layers of the stems and, together with the descending wick-like branches, these facilitate the rapid movement of water up the plant to replace that lost by evaporation. The process also carries minute amounts of mineral nutrients from the water at the base of the plant to the growing points—thus enhancing survival even in very nutrient-poor habitats. Empty hyaline cells can very rapidly fill with water and this gives *Sphagnum* a sponge-like property. Mosses reproduce by spores and in *Sphagnum* species these are produced in small, green globose capsules borne on short stalks that arise near the apex of the plant. The ripe capsules turn glossy brown as the spores mature. Finally, the capsule lid and a mass of spores are shot into the air. This is the result of tension caused by the uneven drying of tissues and increased air pressure in the capsule.

All species of *Sphagnum* are found in acidic habitats and most are very tolerant of a low nutrient status. Some grow in open water where, by a process comparable to that of ion-exchange resins used in industry, the mosses further increase the acidity. Few other plant species are able to put up with such acidic, waterlogged habitats. As a result, *Sphagnum* species are particularly frequent in the high rainfall areas of northern and western Britain. In many regions boggy ground is referred to as 'moss', a place where bog mosses dominate the vegetation. Certain rare flowering plants such as wild cranberry and bog rosemary colonise *Sphagnum* cushions and their distribution in Britain mirrors that of the healthy growth of certain species of *Sphagnum*. The bog mosses are also important as they are often the first species involved in the succession from open water to a drier moorland or woodland community.

Bog moss showing branches and spore-containing capsules

Most of the fungal and bacterial species present in large numbers in drier, more neutral soils are absent from the waterlogged, acidic habitats colonised by species of *Sphagnum*. As a result the dead remains of the bog mosses, and other plants of acidic bog land such as the cotton grasses and heathers, do not decay but the organic matter builds up over hundreds of years to form peat. The fenland peat of low-lying parts of eastern England is mostly composed of the undecayed remains of reeds and sedges. Peat is combustible when dry and has provided a valuable domestic fuel, especially in parts of Scotland, Ireland and Wales.

I still possess a rather strange-shaped spade from the time I lived in Ireland where I rented a section of peat bog in much the same way that I now rent an allotment. During the early part of the summer I dug hundreds of heavy brick-shaped sods from my cutting. These were laid out to dry and then stacked in little 'castles' for a further period of drying before I carted them back to keep my home fire burning. The characteristic smell of burning peat is put to good use in the peat fires used to dry malt in the production of whiskey. My peat cutting was only a few miles from the village of Bushmills, well known as the home of a very fine malt whiskey. The distillery in Bushmills proudly claims to be the world's oldest whiskey distillery: in 2008 it will celebrate 400 years of production and I look forward to downing a glass of Black Bush to mark such an important milestone.

In the south of Ireland peat has been used to fuel electricity-generating stations, although here the digging and collection of the peat are highly mechanised. Mechanisation has also stripped large areas of peat from English deposits in the Somerset Levels and in the Thorne Moor area near Doncaster. This peat supplies the demands of gardeners and the horticultural trade where it is used as a growing medium and to make pots. Conservationists have grown increasingly worried at the habitat destruction resulting from the mass extraction of peat from lowland sites—hence pleas from the Green movement for people to use alternatives to peat-based composts.

My peat spade has a very sharp flat blade (30cm long, 12cm broad) with a small spur-shaped blade projecting at right-angles from its base. Just such a dangerous weapon caused a serious injury to a German peat-cutter in the late 19th century. His arm was badly cut by a spade and his colleagues wrapped the wound with fresh *Sphagnum*. (The use of bog moss as a wound dressing is a common feature in the herbal folklore of many parts of northern Europe.) Ten days later the man was treated at a hospital in Kiel. On removal of the moss the wound was found to have almost totally healed. This sparked an interest by Professor Neuber of the Kiel Clinic, who later extolled the virtues of *Sphagnum* in a series of scientific papers.

Over the following 30 years bog moss was extensively used in German hospitals. After its successful use by the Germans in the early months of World War I it was formally approved as a wound dressing by the British War Office in 1915. Large quantities of the moss were collected, dried, placed in bags of butter muslin, sterilised and sent to the front. The process was repeated in World War II as supplies of cotton-wool dwindled. As a dressing *Sphagnum* has twice the absorptive power of cotton-wool, and its constituent fungal and other micro-organisms give it antiseptic and antibiotic properties. These properties help to explain another former use for *Sphagnum*, as an effective nappy. Perhaps a green nappy would help to ease the huge landfill problem caused by modern disposable ones.

Cushion-like growth of Sphagnum

Moss

'Patents' will burn it out; it would lie there
Turning white. It shelters on the soil; it quilts it.
So persons lie over it; but look closely:
The thick, short green threads quiver like an animal
As a fungoid quivers between that and vegetable:
A mushroom's flesh with the texture and consistency of a kidney.
Moss is soft as a pouch.

Extract from *Moss* by Jon Silkin

Coltsfoot

Tussilago farfara

Coltsfoot is widely distributed throughout Britain, although it was not recorded from the Isles of Scilly until the beginning of the 21st century when I noticed it growing outside the guest house where I was staying! It is a plant of open and disturbed ground including river banks, railway tracks, gravel pits, colliery spoil heaps, limestone quarries, coastal boulder-clay cliffs, pavements and car parks. Many years ago I picked some coltsfoot leaves from the edge of a car park where I had just changed a flat tyre. To my delight the crumpled leaves quickly removed the mud and grease; I had chanced on a natural Swarfega.

Coltsfoot leaves contain a range of chemicals including mucilages, tannins, inulin, a bitter glycoside and saponins. My clean hands were the result of the lather-producing saponins. For at least 2000 years *Tussilago* has been used not as a hand cleaner but as a cough dispeller. Coughwort is an old English name for the plant and *tussus* was the Ancient Greek word for a cough. Pliny (AD 77) recommended coltsfoot's medicinal virtues: 'The smoke of this plant, dried with the root and burnt is said to cure, if inhaled deeply through a reed, an inveterate cough'. Many 19th century Parisian pharmacies depicted a painting of coltsfoot on their doors and coltsfoot still features in Chinese medicine where the cough medicine 'kuan dong hua' is made from the dried flower heads of *Tussilago*.

That smoking might cure a cough is anathema to modern medical practitioners but the expectorant, antitussive and demulcent (soothing) properties of a herbal cigarette made from coltsfoot, and other dried plants such as yarrow, eyebright (page 45) and chamomile (page 101), put it in a different league to tobacco. The smoke is also antispasmodic and relaxes the openings of the bronchial tubes in the lungs. This helps to explain why coltsfoot cigarettes were recommended for 'shortness of breath' and are still occasionally used in the alleviation of asthma symptoms. The leaves were also formerly used in the production of special 'pectoral' beers sold for the treatment of chest complaints.

For hundreds of years coltsfoot leaves were smoked as a cheap alternative to the 'obnoxious weed', a fact remembered in yet another common name—old men's baccy. During World War II, when tobacco was strictly rationed, coltsfoot leaves were once again in demand. For non-smokers a tisane or syrup extract made from the leaves (or flower heads) is also effective in the treatment of pulmonary infections including whooping cough and bronchitis. Coltsfoot leaves were removed from the British Pharmaceutical Codex in 1949, but coltsfoot is still a common component of many herbal cough medicines. I have met retired miners in the Yorkshire area who recounted how they drank coltsfoot tea to soothe the coughing exacerbated by coal dust.

The bitter glycoside found in the leaves stimulates the flow of bile in a manner similar to the earlier use of wormwood (see page 131). In a cough mixture, coltsfoot leaf extract is usually mixed with honey or liquorice to mask its bitter taste. The leaves were formerly used in poultices

Coltsfoot flower heads with central disc florets and peripheral ray florets

to alleviate the results of insect bites, ulcerative wounds and piles. The high zinc and mucilage content of the leaves is the likely healing component. Coltsfoot root boiled with sugar or honey is still used in the preparation of coltsfoot rock, a cure for coughs and sore throats. As it was not classed as a sweet, the rock escaped rationing in the years after World War II and this helped to boost its sales to children.

Other English names for the plant relate to its appearance and unusual annual cycle. Coltsfoot was formerly foal's-foot, a translation of the medieval Latin name *pes pulli*. The leaf blade is indeed shaped like a small (young) horse's hoof and, unusually for a herbaceous plant, it does not unfold until after the flowers. This was recorded in an early description of the plant, *'filius ante patrum'*, translated as another local name, 'the son before the father'. Despite these unusual features several related species have been confused with coltsfoot (see below).

The plant over-winters in the form of a creeping underground rhizome. As early as January (February or March in more northern districts) flowering stalks emerge covered in woolly, purple-green fleshy scales, but no leaves. Each stalk is up to 15cm long and terminates in a single yellow daisy-like flower head (2–4cm across). This is erect in bud, later droops as the flowers wither and finally turns up again in fruit. The golden yellow centre of the flower head consists of many tubular, pollen-bearing disc florets (small flowers) surrounded by several rows of slender, sulphur-yellow, petal-like ray florets, each with a forked stigma. In contrast, early flowering dandelion plants have a smooth stem bearing a solitary flower head consisting only of ray florets. Coltsfoot flower heads open in sunny weather but close in dull conditions and at night. The nectar-rich flowers are typically pollinated by flies and other insects.

Each seed is surrounded by a ring of white hairs (the pappus) that acts as a very effective parachute mechanism. Single seeds have been tracked over a distance of 4 km 'blowing in the wind'. Dispersal is aided by a very noticeable extension of the flower stalk (to 30cm tall) before fruit maturation. This results in a higher launchpad for the seeds. It is only after the flowers are over that the leaves emerge. This caused Pliny to surmise that coltsfoot did not have leaves, whereas both Gerard (1597) and Parkinson (1640) included two illustrations labelled 'Tussilago florens' and 'Tussilaginis folia' for the flowering and leaf stage, respectively.

*Coltsfoot leaves—
note white felty
underside*

Each leaf arises from the root stock (not from a stem) and the blade is carried by a long, grooved, purple-tinged petiole. The hoof-shaped broad blade (10–20cm across) only develops fully by mid-summer. It has a heart-shaped base, pointed tip and scalloped margin, as if bitten into, edged with triangular teeth. Young leaves are covered with a white felty down on both sides but this soon washes off from the upper surface. The leaf outline is comparable to that of the human lungs and some authors quote coltsfoot as an example of the Doctrine of Signatures, where the shape of part of a plant provided a clue as to its medicinal properties.

In Britain there is another plant in the daisy family where the flowers arise before the leaves—Gerard declaring 'Butter Burre doth bring foorth flowers before the leaves as Coltsfoot doth'. As a result it too was previously known in some districts as 'the son before the father'. Although the flower heads of butterbur are produced from a thick scaly stalk, the flower heads are not solitary but clustered round the upper part of the stem. The florets are all tubular and of a lilac-pink colour. In most parts of Britain only female plants are present, although in the north Midlands and Peak District both sexes occur and masses of feathery seed are set. The leaf of butterbur is felty-grey only on the underside and more rounded in outline than that of coltsfoot. By the end of the summer the blade may attain a width of one metre, much bigger than that of coltsfoot, resulting in the name wild rhubarb. Butterbur is typically found on wet ground, often by rivers and streams where, in the days before refrigeration, the leaves were used as a wrapping for butter placed in the water to keep it fresh.

Coltsfoot was also used in a number of more obscure ways. The white felty covering from the leaves was scraped off, soaked in saltpetre and dried, making excellent tinder for the starting of fires in the days before matches. The hairy seeds were once collected in parts of Scotland and stuffed into mattresses and pillows. Pig farmers fed the leaves (they called it hogweed) to their animals in the belief that it brought a shine to a pig's hide. Finally, the flowers can be made into a rather good white wine. From the products of this wonderful plant it is possible to keep healthy while having a smoke together with a glass of beer or wine and not owe any duty to the government: what bliss!

Extract from *Weeds*

Some people are flower lovers
I'm a weed lover.

Weeds don't need planting in well drained soil;
They don't ask for fertilizer or bits of rag to scare the birds
They come without invitation;
And they don't take the hint when you want them to go.
Weeds are nobody's guests:
More like squatters.

Coltsfoot laying claim to every new dug clump of clay;
Pearlwort scraping up a living between bricks from a ha'porth of mortar.
Dandelion you daren't pick or you know what will happen;
Sour-docks that make a first rate poultice for nettle stings
And flat-foot plantain in the back street, gathering more dust than the dustmen.

Even the names are a folksong:
Fat hen, rat's tail, cat's ear, old men's baccy and stinking billy
Ring a prettier chime for me than honeysuckle and jasmine.
And sweet cicely smells cleaner than sweet william though she's barred from the garden.

Norman Nicholson

Emerging flower head of butterbur

Deadly Nightshade

Atropa belladonna

The family Solanaceae contains some very important edible species including potato and tomato, but it is also home to many highly poisonous plants, of which the most infamous is deadly nightshade. The name is better known than the plant, which many people confuse with the much more common woody nightshade (*Solanum dulcamara*). This is despite the fact that the two species are very different in habitat and growth habit, not to mention leaf shape or fruit size and colour! Woody nightshade is a perennial woody climber, typically seen in hedgerows, where its clusters of small, bright purple flowers and small tomato-like fruits stand out from the crowd.

Deadly nightshade is a perennial, non-woody, much-branched plant that grows to about 1.5 metres tall with the appearance of a small shrub. As a wild plant it is most frequent on calcareous soils in the south and east of England, typically on disturbed or waste ground including rabbit burrows and in old quarries. It is also frequently found close to former monastic sites, my latest sighting was at Fountains Abbey, where it is probably a relic of cultivation. Deadly nightshade plants have long been associated with the region known as the 'vale of the deadly nightshade' close to the ruins of Furness Abbey in Lancashire.

The thick, angular, often purple-tinted stems bear large oval, pointed leaves similar to those of tobacco (also in the Solanaceae). The leaves are either alternate or in pairs where one of the pair is much larger than the other. The crushed leaves have a disagreeable smell. The long-stalked, large, tubular bell-shaped flowers are borne singly or in pairs from both the leaf axils and stem junctions in the upper part of the plant. The petals vary in colour from dark green to dull purple or brown. The sweet-tasting fruits look like glossy black cherries (cupped by five green sepals) but, along with the roots, stems and leaves, are extremely poisonous.

Atropa belladonna is now rarely grown in gardens and where present it is normally kept away from places where children, and others, might be tempted to eat its fruit. One exception to this is the new poison garden at Alnwick Castle where the Duchess of Northumberland has included *Atropa* alongside some other potent plants. As early as 1597 Gerard warned, 'banish it from your gardens and the use of it also, being a plant so furious and deadly: for it bringeth such as have eaten thereof into a dead sleepe wherin many have died …'. By Victorian times the message was much the same: 'It is a plant that we should gladly see growing in our own garden, but we should dread the responsibility of putting it there, for, if our younger children came to harm through its temptations, we should expect a jury to return a verdict of constructive homicide.' (*Familiar Wild Flowers*, Hulme, undated).

The plant's poisonous nature is alluded to in its generic name, taken from Atropus, one of the three Fates in Greek mythology who held the shears that could cut the thread of life. Belladonna is a former common term for the plant that was incorporated into its scientific name. A 16th-century reference mentions Venetian ladies using *Herb bella donna* (beautiful lady herb)

*Bell-like flower of
deadly nightshade*

in a cosmetic to enlarge their pupils. Men supposedly found the wide-eyed look alluring, or it may just have been that the women became too short sighted to see what they were getting!

Consumption of just two or three of the fruits can kill a child, although rabbits and goats are not poisoned, and the flea beetle *Epithryx atropa* can eat the foliage without ill effect—one man's poison is another animal's vegetable. The chemicals isolated from *Atropa* include a number of alkaloids such as hyoscyamine, scopolamine and atropine, the latter named after the plant but actually only produced by conversion from hyoscyamine when the plant is dried. Atropine and hyoscyamine inhibit the effect of acetylcholine, a neurotransmitter in the human autonomic nervous system. Some of the alkaloids' effects are summed up by the old adage describing a patient with deadly nightshade poisoning as 'hot as a hare, blind as a bat, dry as a bone, red as a beet and mad as a hen'. In more scientific terms the dilation of the subcutaneous blood vessels and pupils results in flushed skin and blurred vision. The drying up of secretions, including saliva, results in a dry mouth, while hallucinations give the appearance of madness (although I cannot see the link between hens and madness).

Other symptoms include a decrease in gut movement and an increase in heart rate coupled with the heartbeat becoming clearly audible. A good GP is said to be able to detect deadly nightshade poisoning by listening to the patient! Lethargy and a period of deep sleep (Gerard called the plant sleepy nightshade) are common symptoms associated with high doses that may result in coma and death. The alkaloids extracted from *Atropa* can be absorbed through the skin; knowledge used by generations of witches who smeared their bodies with concoctions made from deadly nightshade and other plants rich in alkaloids such as henbane and hemlock. The resulting sleep was reputedly filled with vivid dreams and hallucinations, including the feeling of flying through the air—the archetypal trip. Witches have frequently been depicted flying through the air on a broomstick, the shaft of which was used to rub the 'flying ointment' into the vagina from where the alkaloids passed into the blood stream. This neatly bypassed the digestive system, thus limiting symptoms of sickness and diarrhoea. Harry Potter is unable to use this method, so we must seek another explanation for his means of levitation in the broomstick game of quidditch.

Alcohol infused with deadly nightshade was credited in a 16th century *History of Scotland* as having induced stupor in an army of invading Danes, who were thus easily overpowered and killed by the Scottish soldiers led by Macbeth. In Shakespeare's *Macbeth*, after meeting the witches, Banquo asks 'Or have we eaten on the insane root, that takes the reason prisoner'— possibly deadly nightshade; a common ingredient of a witch's brew. Later, Edwardian England was rocked by the antics of Dr Crippen, who used hyoscyamine extracted from henbane, a close relative of deadly nightshade, to murder his wife before fleeing to Canada with his mistress.

As with so many poisonous plants (see foxglove, pages 50–53), deadly nightshade has a long history of being used medicinally. An extract from the dried leaves and roots was applied to the skin as 'sorcerer's pomade' where it knocked patients out before operations were performed. A major problem was that of dose, as the concentration of the alkaloids varies in different parts of a plant, in different seasons, from plant to plant and with the state of preservation of the dried plant material.

Once the alkaloids were isolated from the plant the problem of dose was easier to control. Deadly nightshade is still grown in parts of southern Europe for the pharmaceutical trade. Mention has already been made of *Atropa* being cultivated in England by monks, probably for medicinal reasons, and 16th-century herbals record that it was being grown in limited amounts in Britain. The Balkan and First World Wars curtailed imports to Britain, thus helping our local suppliers; the plant was still being cultivated in parts of Suffolk and Hertfordshire towards the end of the 1940s. Atropine is now obtained commercially from Australian plants of the genus *Duboisia*.

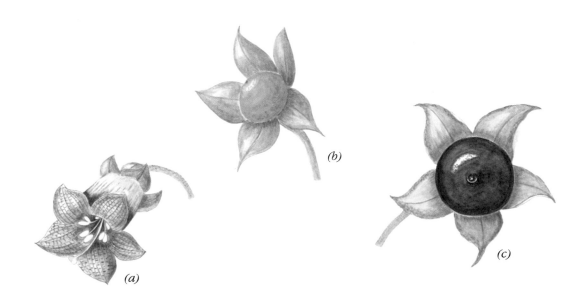

Deadly nightshade   (a) Flower   (b) Immature fruit   (c) Mature fruit

The symptoms of deadly nightshade poisoning are very similar to those of scarlet fever. Homoeopathic medicine seeks to treat like with like, that is to use a very small amount of a substance to treat an illness with similar symptoms to that produced by a large dose of the substance. As such, belladona is commonly used by homoeopaths in the treatment of scarlet fever and for headaches. In *Little Women* (1868) Beth is advised to 'go home and take belladona right away' when she is found to have been nursing a baby who had died of scarlet fever. It must be stressed that self-administration of *Atropa* is very dangerous; it can kill as easily as it can cure.

Mainstream medicine still makes use of the alkaloids found in *Atropa*, although some are now synthesised or have been replaced by more modern drugs. Before ophthalmic diagnosis or surgery a tiny dose of atropine infused on a disc of gelatine is placed over the eye. The ensuing dilation of the pupil assists with the examination or surgery. Some travel sickness drugs and remedies for stomach ulcers incorporate atropine and it can also be effective in the treatment of stomach cramps and irritable bowel syndrome. Atropine, like the foxglove derivatives (see page 52), is sometimes used to treat myocardial infarction (abnormal heart beat) following general anaesthetic or a heart attack.

Most importantly, atropine is still administered as part of the premedication before a general anaesthetic. The resultant reduction in the production of saliva makes the job of the anaesthetist much easier as too much saliva can result in the patient choking. The reduction in bodily secretions initiated by atropine has proved effective in the treatment of peptic ulcers. It also eases muscle rigidity and its mode of action has paved the way for modern synthetic drugs used in the treatment of Parkinson's disease. The general public has little idea as to how much debt we still owe to our medicinal plants!

Woody nightshade growing in a hedge

Dove Tree

Davidia involucrata

In contrast to the story behind ivy-leaved toadflax (see pages 58–61) the introduction of the dove tree to British gardens is well recorded, largely because it is a much more recent addition to our garden flora. To appreciate fully the full story of *Davidia* we need to set it against a background of history, plant politics and cross-channel rivalry.

The dove tree is aptly named after its likeness, when in flower, to a tree festooned with doves. It is also known as the ghost tree, handkerchief tree and, rather more prosaically, Kleenex tree. *Davidia* is native to China, a country that for ages proved inaccessible to most Europeans, including professional plant hunters.

In the 16th century Western trade with China was monopolised by the Portuguese and importation of garden plants was virtually non-existent. In the 17th and early 18th centuries Jesuit missionaries went to China, but even those interested in plants were largely thwarted in their attempts to introduce species by the long journey time. For the British any trade with China was under the control of the East India Company. Following cessation of the Company's monopoly and the end of the Opium Wars (1839–1842) trade with China was opened up. This was largely due to Lord Palmerston's successful negotiations culminating in the Treaty of Nanking (1842) initiating the lease of Hong Kong as a base for trade links with mainland China.

Although visitors were initially restricted to certain Chinese ports along with a narrow coastal strip, the Horticultural Society, yet to receive its Royal patronage, sent Robert Fortune to China, via Hong Kong, in 1843. Fortune introduced many new plants to Europe—most famously tea. The restriction of foreigners to the coastal regions meant that Fortune was only scratching the surface of the enormous floristic diversity that was China's legacy.

Fortunately for Western gardens several outstanding European naturalists were to find themselves working in China's interior during the latter half of the 19th century. In 1862, just months after Fortune's last visit to China, a French priest, Père Jean Pierre Amand David, was sent as a teacher to a Lazarist (Catholic) Mission in Peking. In his spare time David collected plant, animal and geological specimens and sent them to the Paris Musèe d'Histoire Naturelle (next to the Jardin de Plantes). The museum's director was so pleased with the specimens that he persuaded the Lazarists to let David have more time to explore China's natural history.

During three extensive trips he collected specimens of over 2,000 species of plant, as well as describing animals such as the giant panda. In 1868–69 his excursions took him to the Tibetan border and in the Mupin region, close to Tibet, he was probably one of the first westerners to see the dove tree. He sent back herbarium sheets of the flowering twigs to the Paris museum, but was too early in the season for seed collection. The dried specimens caused a sensation, but the plant remained a fascinating curiosity in the West as David was forced to retire in 1874 through ill-health. David died in 1900, just as the story of the plant named in his honour was reaching a crescendo.

A second French missionary, Père Jean Marie Delavay, was sent to China in 1867 and remained there until his death in 1895. Over nearly 30 years he collected seeds and over 200,000 plant specimens that he sent to the Jardin de Plantes. Sadly, *Davidia* was not among them. The third naturalist was Dr Augustine Henry. Born in County Antrim, now part of Ulster, he qualified as a doctor in Edinburgh and in 1881 was appointed as medical officer to the Imperial Chinese maritime customs service, a post he was to hold for nearly 20 years. Henry was stationed not at a coastal port, but at Ichang in central China, nearly 1000 miles up the Yangtze River, on which boats travelled all the way from the China Sea.

Henry was a man of many talents and became deeply interested in the local flora. He acted as a part-time plant collector and sent over 150,000 herbarium specimens to the Botanic Gardens at Kew and Edinburgh. These included the flowers and leaves from a single specimen of *Davidia* that Henry discovered in Hupeh in 1888, about 1,000 miles to the east of David's original find made some 19 years earlier. Being unable to wait for the fruit to develop, Henry could not supply seed to Kew.

Henry's herbarium specimen persuaded Sir Harry Veitch, who ran the famous Chelsea-based nursery, to fund a collector to bring back seed of the dove tree. The man chosen for the job was Ernest Henry Wilson. Wilson was born in Chipping Camden (where there is a memorial garden planted with his introductions) and worked his way up via the Birmingham Botanic Gardens to a job in Kew Gardens. In 1899, at the age of 23, Wilson arrived in Hong Kong. By then Père David was in his last year and the precise location of his *Davidia* find was uncertain, so Wilson set out to meet up with Dr Henry. To make matters more difficult Henry had been

Dove tree with immature fruit

Leaf underside of dove tree

relocated to Simao in the Yunnan province of south-west China, 1,000 miles from Hong Kong.

China at the beginning of the 20th century was not an easy place for foreigners, not least with the beginnings of the Boxer Rebellion. Wilson's trip to Simao, where Henry now worked, was packed with incident and took over 3 months. Henry provided Wilson with details of the whereabouts of the tree that he had seen some 12 years earlier by drawing a rough map on half a page torn from a notebook. The map approximated to some 20,000 square miles!

Wilson returned to Hong Kong and then travelled up the Yangtze, arriving in Ichang on 24 February 1900. In mid-April he set off by houseboat to Badong, from where his party continued to the house where Henry had stayed close to where he had found the *Davidia*. Finally, on 25 April, with the help of local residents who remembered Henry's visit, he was guided to the tree. All that remained was a stump next to a new house sporting a roof made from *Davidia* timber!

Wilson vowed that he would have to seek out David's original tree but remained collecting a range of plants in the Ichang region. On 19 May he came across a *Davidia* in full flower and a little later he found another twenty trees. It was nearly six months before he was able to collect sufficient seed, which was shipped back to London. The seeds arrived at Veitch's Nursery early in 1901, but by the time Wilson returned in April 1902 none had germinated. Fortunately, many of the seeds that had been planted outside started to germinate in May (although very few of those placed in glasshouses produced plants). The first of these plants flowered in 1911, but sadly for both Wilson and Veitch it was not the first *Davidia* to flower in Europe.

Unknown to Veitch, a Roman Catholic missionary, Père Paul Farges, had been in the Ichang area some three years before Wilson's visit. Farges had sent his seed collection to Vilmorin's Nursery in France and this included 37 seeds of *Davidia*. Only one germinated, in 1899, and this tree flowered for the first time in May 1906. Oddly, both the French tree and those from Wilson's collection had leaves with a smooth undersurface. These differed significantly from Père David's original herbarium specimen. This had soft white down on the leaf undersurface and is now referred to as *Davidia involucrata var. involucrata*. Wilson was to collect seed of this form on his second visit to China, but the form originally collected both by Wilson and Farges (*Davidia involucrata var. vilmoriniana*) is faster growing and outnumbers *var. involucrata* in Britain by at least 5:1. Recent research has shown that the forms have different chromosome numbers and it is possible that in future they will be reclassified as separate species.

When not in flower the tree's broad heart-shaped toothed leaves can easily be mistaken for either a lime or a mulberry. In May the long-stalked, pendulous flower heads develop. Each is a 2cm wide brush-like mass of stamens enclosing a central stigma and ovary. Surrounding the rather insignificant flower head are two pendulous, creamy-white involucral bracts (modified leaves surrounding the flower head, hence the Latin name), the larger some 15cm long and about twice the size of the other. These result in the dove or handkerchief appearance of a flowering tree and draw large crowds to gardens in the late spring.

As a footnote to the dove tree story, Augustine Henry finally returned to his native Ireland where he held the post of Professor of Forestry at the Dublin College of Science. He died in 1930, as coincidentally did Ernest Wilson, who was by then director of the Arnold Arboretum in America. Following countless adventures and the introduction to both Britain and America of many hundreds of plant species, Wilson was killed in a road accident when his car skidded on, of all things, some wet leaves.

Detail from Wilson's herbarium sheet of dove tree

Winter Trees

The wet dawn inks are doing their blue dissolve.
On their blotter of fog the trees
Seem a botanical drawing—
Memories growing, ring on ring,
A series of weddings.

Knowing neither abortions nor bitchery,
Truer than women,
They seed so effortlessly!
Tasting the winds, that are footless,
Waist-deep in history—

Full of wings, otherworldliness.
In this, they are Ledas.
O mother of leaves and sweetness
Who are these pietas?
The shadows of ringdoves chanting, but easing
nothing.

Sylvia Plath

Elder

Sambucus nigra

In a manner comparable to that of juniper (see pages 62–65), elder is variously described as a shrub or small tree. Not only is it frequently less than 6 metres tall (the minimum height requirement to be accepted as a tree) but it usually branches from just above the ground, so purists class it as a shrub and banish it from books about trees! Grigson described it as 'neither bush nor tree, neither bad entirely nor entirely good'. The trunk diameter rarely exceeds 50cm, so it has little value as a timber tree, but just about every part of the plant has been put to good use. In folklore elder has always been a sinister tree associated with death and witches (who were thought capable of hiding from pursuers by transforming themselves into an elder); it is a strange mixture of good and evil.

A common native plant in Britain, although rarer in Scotland, elder typically grows in woods and hedgerows and near human habitation. It prefers damp soil and disturbed sites such as rabbit warrens, badger sets, rubbish tips and previously cultivated ground, especially where the soil is rich in nitrogen. It grows quickly from seed but is short lived compared with other trees and shrubs. The smooth young green twigs age to grey and are filled with a spongy white pith. On older plants the bark becomes deeply fissured. The winter buds are borne in opposite pairs and are unusual in that they lack any protective scales. The purple-green leaf buds open as early as February and the autumn leaf fall occurs as late as November, so the tree is only 'naked' for a very short winter period.

Elder bears compound leaves made up of five (occasionally three or seven) oblong, toothed leaflets. The leaves are not unlike those of ash or mountain ash but are easily distinguished on account of the obnoxious 'cat urine' smell emanating from them when crushed. This helps to explain a former common name for the plant—God's stinking tree. The religious element is derived from the medieval myth that elder provided the timber for the cross of Christ. Tiny five-petalled, sickly sweet-scented flowers are produced in upright, flat-topped clusters during June and July. During late summer and autumn the small globose fruits ripen from green to shiny black (hence the Latin tag *nigra*) and the previously erect clusters droop under the weight of fruit.

Three other species of elder are to be found in Britain, including the rare, dwarf elder (*Sambucus ebulus*), a non-woody plant found on rough ground and in woodland clearings. The red-berried elder (*Sambucus racemosa*) has become naturalised in both northern England and Scotland. It differs in having creamy-yellow flowers, bright red (occasionally yellow) mature fruits and stems containing brown pith. The American elder (*Sambucus canadensis*) is occasionally naturalised; it has larger leaves with nine leaflets and purple-black fruits. Garden cultivars of common elder include 'Aurea' with golden yellow leaves and the increasingly popular 'Black Lace' which has dark purple, deeply cut leaves and pink flowers.

Flower head of elder

The scientific name *Sambucus* probably derives from sambuca, an early Greek musical instrument, but other similarly named pipe instruments (such as the Italian pipe known as the sampogna) may also have contributed to the name. In Britain the stems have been turned into simple panpipes and penny whistles, and also toy blowguns to shoot haws, dried peas or even elder berries, as alluded to in Shakespeare's *King Henry V*: 'That's a perilous shot out of an elder gun'. The ease of removing the pith gave rise to another local name 'bore' (or 'bour') tree. The name elder comes from the Anglo-Saxon *eldrun*, from the word *aeld* meaning fire. This alludes to use of the hollowed stem for blowing air into a fire in a manner similar to the use of bellows.

Elder pith has also been put to a variety of uses. Students of botany are required to cut thin sections for anatomical study and elder pith provides support for flimsy material such as leaves without blunting the razor. Early experiments with electricity made use of the insulating properties of elder pith. In the 1990s, just before a watch repairer retired from his shop near my Sheffield home, I enquired of its use in his trade. He told me that as an apprentice it had been a standing joke when a watch was brought in for cleaning to ask someone to pass the pith pot (try saying it quickly). The elder pith was used to clean dirt and excess oil from the delicate mechanism. These days Blu-tac substitutes for elder pith.

It has long been deemed unlucky to have elder in the house and even to burn the wood (which gives more smoke than flame). Hearse drivers in the days of horse power used whips with elder wood handles and of the many folk tales involving elder none is more ubiquitous than the belief that Judas Iscariot hanged himself from an elder. Shakespeare took up the tale, in his best punning tradition, in *Love's Labour's Lost,* Act V, Scene II:

Dumaine:	*Judas Maccabaeus clipt, is plain Judas.*
Biron:	*A kissing traitor.—How art thou prov'd Judas?*
Holofernes:	*'Judas I am'—*
Dumaine:	*The more shame for you, Judas.*
Holofernes:	*What mean you, sir?*
Boyet:	*To make Judas hang himself.*
Holofernes:	*Begin, sir; you are my elder.*
Biron:	*Well followed: Judas was hanged on an elder.*

Part of the story still survives (or is it part of Judas?) with the strange brown jelly fungus that commonly grows on the stems of elder elder (rather than on young elder). It is known as Jew's ear fungus (*Auricularia auricula-judae*). Some politically correct authors are now calling it juicy ear fungus, but the Jewish link is an important part of the story. When wet it is indeed like a very supple ear and has long been used to flavour 'mushroom' soups. It was formerly used as a palliative for sore throats.

The medicinal value of elder is not restricted to the fungus that frequently adorns its stems. To the herbalist elder provides a wide range of remedies and has been described as a complete medicine chest. An infusion of the dried bark was previously used for its purgative properties in treating constipation. It is also diuretic and was used in the treatment of liver and kidney disorders. Salves prepared from the simmered leaves of elder combined with beeswax and olive or linseed oil speed the healing of sprains, bruises, sore nipples and haemorrhoids.

Elderflower water was used in the 18th century to 'remove freckles' and is still used to alleviate the pain of sunburn. Anti-inflammatory chemicals extracted from the flowers are used in skin creams and for reducing the pain caused by chilblains. An infusion of the flowers promotes sweating and is prescribed by herbalists in the early stages of feverish colds and influenza. The ripe berries are rich in vitamins A and C. Made into a syrup, the fruits serve as a prophylactic against colds and flu.

The flowers and fruits of elder are as at home in the kitchen as they are in the medicine cabinet. Small sections of the flower head dipped in batter, deep fried and eaten with a dusting of sugar make an unusual summer pudding. The flower heads are easily and quickly made into cordial (or presse as the more expensive brands proclaim themselves) and over a slightly longer process into the gloriously effervescent alcoholic elderflower champagne—although following complaints by the French it can no longer be marketed as such! The berries make fruit jellies and pickles. They were formerly used in the production of fraudulent port and still make a good home-made wine, which can be drunk young as a mulled Christmas drink.

Other uses for the plant include the tying of its leaves to the harness of a horse to keep flies away. Crushed elder leaves were also worn on hats to keep mosquitoes at bay. The wood can be made into skewers, pegs and combs; the timber rots very slowly and was used for slender fence posts, when it was said to last longer than iron. The bark and roots yield a black dye once used for colouring wool. An extract from the leaves when added to alum or chrome makes a rich green dye while the berries provide a deep violet colour. It was indeed a plant to dye for as well as a plant associated with death!

Elder is associated with an ancient Celtic 13-month calendar, of which the old Gaelic version is known as Beth-Luis-Nion (Birch-Rowan-Ash). Each 28-day month was named after a tree that had a particular place in the year due to symbolic and magical links. Elder, or Ruis as it was called, gave its name to the final month from 25 November to 22 December. The first month (Birch) ran from 24 December and the thirteen 28-day months totalled 364 days, with Yew (see page 134) standing for 23 December and thus completing the typical solar year.

Extract from *The Tree Calendar*
Elder   Ruis

Elder in its own season—
the old dregs of the year—
is utterly empty, surely the worst
tree in the hedge, stripped dead,
gnarled and twisted afflicted
sticks, like some crabbed arthritic
witch, pale driftwood blown
inland by the gale, to catch
grotesquely by its twigs: only force
of will can colour it now thick with fruit,
or remoter still, cream-flowered,
smelling of days past, days of promise, summer.

Hilary Llewellyn-Williams

Shiny black berries of elder

Ergot

Claviceps purpurea

It is 40 years since my days as an undergraduate in the mid-1960s, but I can still picture the puzzled looks on the faces of the good people of Bangor as the letters LSD began to appear among the local graffiti. It is many millennia since our ancestors settled down from a hunter-gatherer lifestyle to one in which the cultivation of cereals provided our daily bread. The link between these two statements is a visually insignificant fungus known as ergot. The fungus has left its mark on civilisation out of all proportion to its benign appearance.

The old French word *argot* means a cockerel's spur and this has given us the name for a fungus that also resembles a tiny, hard, purple-black banana, up to 2cm long and just 2 mm in diameter. Ergot is parasitic on the flowers of members of the grass family, including cereals such as wheat, barley, oats and rye; species that were selected for their large grains and have become staple food crops across Europe and other regions of the world. The black structure is a sclerotium (resting structure) and is found in place of a seed or grain of a grass or cereal. In cereals the resultant loss of grain yield is exacerbated by the fact that the fungus also causes sterility in the surrounding flowers of the inflorescence. Unfortunately, this loss of yield is nothing compared to other problems resulting from ergot infestation.

Unlike most of the larger fungi, ergot is classed as an Ascomycete as it produces sexual spores in an ascus (sack-like structure). In infected grasses and cereals the sclerotia fall to the ground in the autumn where they remain dormant until the following spring. Each sclerotium then produces up to a dozen tiny spore-bearing structures resembling minute pink-headed drumsticks. Spores are discharged from these into the air, where wind currents carry them to the feathery stigmas of nearby grass and cereal flowers. A germinating spore produces a tube that invades the ovary from where resultant fungal tissue produces masses of asexual spores in sugar-rich droplets resembling nectar. These are dispersed to surrounding flowers both by rain splash and by insects attracted to the sweet solution. By late summer the infected grass/cereal ovaries have produced new sclerotia—the life cycle is complete.

The late summer harvest of cereal crops and the subsequent thrashing out of the grain, together with sclerotia from infected plants, interrupted the fungal life cycle and unleashed a poisoned chalice on those eating the fruits (usually in the form of bread) of their labour. Medieval records indicate that in extreme cases as much as one-third of the grain harvested in parts of Europe was ergot and not cereal seed. The resultant grey flour was scorned by the rich, who could afford supplies of a better quality or grain that had been sorted to remove 'imperfections'. So it was the poor who suffered more from what we now call ergotism.

An Assyrian tablet from the 6th century BC describes a 'noxious pustule in the ear of grain'. The earliest documented description of symptoms that we now assume were caused by the eating of ergot-contaminated grain comes from Germany in AD 857: 'A great plague of swollen blisters consumed the people by a loathsome rot, so that their limbs were loosened and fell off before death'. In the following century victims of a 'plague of fire' in the Paris region found relief when cared for at a local church—where the bread fed to the patients was made from flour of a superior quality. Later victims of the burning sensation which became known as *ignis sacer* (holy fire) were considered to be suffering as a result of divine retribution and sought help from the Church. A Viennese hospital dedicated to St Anthony successfully treated many sufferers and in 1093 the Order of St Anthony was established. This was followed by the building of churches dedicated to the saint, to whom sufferers prayed for redemption. From this time on the affliction was known as St Anthony's fire, and church frescos were painted showing Anthony surrounded by patients with burning or missing limbs.

Historians have shown that the symptoms of eating infected grain can be separated into two types: gangrenous ergotism and convulsive ergotism. Episodes of the former seem to have been largely restricted to France and areas to the west of the Rhine, whereas convulsive ergotism was found in the rest of Europe, including Britain, and also in North America. Initial symptoms of gangrenous ergotism included the feeling of intense heat, especially near the limb extremities; in more serious cases this was followed by the loss of nails and a numbing of hands and feet which later turned black, dried up and dropped off. This shedding of limbs (and also ears) was largely without pain or much blood loss, but was terrifying for the patient and accompanied by the awful smell of rotting flesh.

Patients suffering convulsive ergotism exhibited sickness, diarrhoea and muscular twitching, together with seizures that resembled epilepsy. They were sometimes diagnosed as having St Vitus' dance, a general term for those with uncoordinated body movements. Many sufferers experienced vivid hallucinations and formication, the sensation of having ants running all over their bodies. The different forms of ergotism may relate to different climate, soil conditions or strains of cereal that alter concentrations

*Ergot   (a) Sclerotia on rye   (b) Spore-bearing sclerotium
(c) Single sclerotium*

of certain toxic chemicals in the fungus. Of all our cereals, rye is the most prone to infection by ergot.

In 1989 Mary Matossian published a book outlining her research into the link between ergotism and the witchcraft trials that took place between the 16th and early 19th centuries. She highlighted the strong correlation between regions with many trials and the rye-growing parts of Europe. Countries such as Ireland, where the carbohydrate crop for the poor was potato, had hardly any. From dendrochronological work she was able to ascertain when regions had been subject to cold, wet springs (years of narrow tree-ring growth). Such conditions extend the flowering season of cereals and favour the spread of fungus spores. She found that poor springs preceded abnormally high incidences of witch trials in the rye-growing regions.

Matossian also investigated the famous Salem witch trials of 1692. Not only did the settlers grow rye (ironically America's native sweetcorn is not susceptible to ergot) but diary records show that in the early 1690s the people of Salem suffered from unusually cold spring weather. Symptoms of the 'bewitched' included pricking sensations, hallucinations and convulsive seizures. Those most affected were the young, who would have been given bread even when their parents went hungry. Children also have a lower body mass than adults and react to a lower dose of the ergot chemicals.

It was not until the 18th century that definite links were made with the ergot fungus and the symptoms described above. By this time rye was a less important crop, and since then cleaning (separation of grain and fungus by flotation), deep ploughing (burying the sclerotia), plant breeding and fungicidal sprays for cereals have almost eliminated ergotism in Europe. The fungus continues to infect grasses and this can result in abortions among grazing animals.

By the 20th century chemists had begun to extract some of the alkaloids found in the ergot sclerotium. One of these, ergometrine, was found to restrict the diameter of blood vessels and act as a uterine stimulant. This confirmed the medicinal use, dating back as early as the 16th century,

Ergot sclerotium on flower head of cocksfoot grass

of an ergot decoction given by midwives to women to induce labour. It was also used to reduce the threat of postpartum haemorrhaging. An injection of ergometrine, or a similar drug (oxytocin), is still used for this latter role today. Another useful ergot alkaloid is ergotamine, which constricts the blood vessels around the skull and has been incorporated into a number of drugs for the treatment of migraine.

By the 1940s various pharmaceutical companies were investigating ergot's other chemicals. Albert Hofmann was working for the Swiss firm Sandoz when in 1943 he examined the properties of one of a range of derivatives produced from an ergot chemical called lysergic acid. He accidentally ingested a tiny amount of D-lysergic diethylamide (LSD) and was the first to experience this very powerful psychoactive drug. By the 1950s the CIA was testing LSD on unsuspecting 'volunteers' in the hope that it might prove useful in the interrogation of spies. Psychiatrists also hoped that the drug would help in the treatment of mental disorders. By the 1960s its use by the CIA and the medical profession was banned and Sandoz stopped production. By then LSD had become part of the anti-establishment scene, initially championed by Huxley's *The Doors of Perception* (1954) and later by a Harvard professor called Timothy Leary. The rest is history.

Talking of history, there is an interesting footnote to the ergot story. For hundreds of years the Ancient Greeks celebrated a form of harvest festival that incorporated the myth of the reunification of Demeter, goddess of agriculture, and her daughter Persephone, who had been kidnapped by Hades. The celebration or the Mysteries of Eleusis took place in Athens. A select few were allowed into the temple of Demeter at Eleusis where they drank *kykeon*, a purple-coloured beer made from meal. Among those experiencing the resultant hallucinations, sweats and tremors were Plato and Homer. Alcohol alone would not have been powerful enough to produce the described effects. Ergot could have been responsible for the beer's purple colour and if so it might have produced many of the 'mysterious' symptoms. *Claviceps purpurea* is a small fungus with a big history.

Different strain of ergot with long, narrow sclerotia

Eyebright

Euphrasia officinalis agg.

The current bible of the dedicated British botanist, *New Flora of the British Isles* (Stace, 1991), lists no fewer than 21 species of *Euphrasia*, adding that they are all known as eyebright. As if there wasn't enough of a problem with over 20 species, this 'highly critical genus' is also blessed with at least 60 identifiable hybrids. Many of these hybrids are fertile, which puts in doubt the designation of the parents as separate species. The fact that some specialists consider that there are far fewer separate species (or microspecies as they are often called) does little to help the beginner. More useful is the fact many of the species have a very restricted geographical range, usually in parts of western or northern Britain. The most widespread species and typically the most common in England and Wales is *Euphrasia nemorosa*, a plant frequently mistaken for the aptly named *Euphrasia confusa*. Other species differ in one or more of the following characteristics: plant size, hairiness, leaf shape, flower size and colour—but they are all recognisable as eyebright!

When, as a child some 50 years ago, I learnt how to identify the wild flowers of my neighbourhood, life was much easier. The book I used, *The Pocket Guide to Wild Flowers*, McClintock and Fitter, 1956, included only one species of eyebright in the main text although a footnote added that it included 24 microspecies! The species given full treatment was *Euphrasia officinalis* agg. (agg. meaning aggregate of a wide range of forms) and the text pointed out that eyebright was very variable, especially in size, branching and hairiness. Taxonomy is an interesting pursuit and any group of organisms can be classified, even botanists. Botanists fall into one of two distinct groups, 'lumpers' and 'splitters'; as one who falls unashamedly in the former category I am happy to enjoy eyebright as a very variable species of beauty and interest without wasting time splitting hairs or the absence of them.

The genus name *Euphrasia* comes from the Greek word *euphraino*, meaning cheerfulness, as immortalised in the name Euphrosyne, the goddess of gladness and one of the three Graces of Greek mythology, who was full of joy and mirth. How does such a tiny plant bring about such happiness? The flowers are beautiful but rarely observed closely enough to appreciate fully their form and colour. The old specific name *officinalis*, indicating that the plant was sold from a storeroom or office, betrays its historic medicinal use. As a medicine the plant is reputed to have brought gladness to all those who partook of it.

Eyebright is a low-growing annual, and except when growing among tall vegetation, is rarely more than 15cm high. Typically the stem (which in some forms is covered in sticky hairs) ranges in colour from green to pink or purple. While some plants consist of a single unbranched stem, in others repeated branching of the main stem gives the plant a shrub-like habit. The small oval, stalkless leaves (about 1cm long) are typically dark green and often develop a purplish-bronze tint. They are hairless on the upper surface and have deeply toothed margins. Mostly in opposite pairs, the leaves are more likely to be alternately arranged towards the apex of the stem.

Honey guides on eyebright flowers

The small flowers (usually less than 1cm across) are produced from the leaf axils in the upper parts of the plant. The petals fuse to form a tube that opens at its apex to form two lips, the upper erect, the lower with three spreading lobes. The petal background colour is typically white but may be lilac or more rarely yellow. Most flowers also have a bright orange-yellow blotch at their centre, this being on the lower lip near the entrance to the throat of the flower. The petals are adorned with a number of radiating purple lines ('honey guides'), which are most frequent on the lower lip.

The eyebrights are species that cover a wide range of less fertile habitats, including dry calcareous grassland, acid heaths and moors, grassy cliffs, old railway tracks and open woodland. Being annuals and regenerating entirely by seed, they require open ground for the successful establishment of their seedlings. Eyebrights are found over a wide altitudinal range and some forms are restricted to upland sites. The larger flowered forms are mostly insect pollinated but many of the smaller ones are self-pollinated. Inbreeding resulting from self-pollination is one of the factors leading to isolated, distinct local forms.

Eyebright is described as a root hemiparasite, meaning that it obtains at least some of its carbohydrates, and possibly water and nutrient salts, from surrounding plants to which it connects via suckers from its root system. Like other green plants, eyebright makes its own sugars using the power of sunlight and can survive on its own, but such plants are typically much smaller than those with a host; one reason for the range of sizes found among eyebrights. Grime et al. (1990) report that the most common host plants are grasses and members of the pea family. The degree of debilitation shown by host plants as a result of being parasitised by *Euphrasia* is unknown.

Earlier reference was made to the medicinal virtues of eyebright. From the 16th century onwards many European herbalists began to incorporate a Chinese philosophy that was to become known as the Doctrine of Signatures. This was the belief that God had left a sign in the shape or colour of part of a plant to indicate its medicinal usage. This is neatly summed up by the phrase 'every plant gave an outward sign of its inward grace'. The swollen tubers of lesser celandine (see page 66) are said to resemble the blistering associated with piles, the plant is still

Eyebright with limestone backdrop

used to lessen the discomfort of haemorrhoids. The appearance of *Euphrasia* flowers, with the central orange-yellow 'pupil' reminiscent of a bruised or bloodshot eye, suggested under the Doctrine of Signatures that the plant should be useful for eye complaints. As with so many of these 'signature plants', eyebright had been used as a medicine long before this theory became fashionable.

Given the Greek origin for its name, it is not surprising that it features in the writings of both Theophrastus and Dioscorides, who prescribed it for eye infections. By the 14th century its use as an eye medicine was mentioned in various European texts and the French came to call it '*casse-lunettes*' (no need for glasses). To early English apothecaries it was known as ophthalmica. The Welsh names include golwyg Crist (Christ's sight). By the 16th century European botanists such as Dodoens and Fuchsius noted it as being used to treat eye disorders.

In the early 17th century the British writer Markham recommended 'Drinke everie morning a small draught of eyebright wine'. Later in the same century there are records of the plant being made into a special ale and in Salmon's *The English Herbal* (1710) he tells us that eyebright did 'strengthen the head, eyes and memory and clear the sight'. Culpeper went even further by asserting 'if the herb was but as much used as it is neglected, it would spoil the spectacle makers' trade'. A later treatise even suggested that the consumption of tea made from eyebright 'will often raise children from the bottom of their classes to the top'. Perhaps we should inform the government that there is a cheap method of improving exam results.

Eyebright contains a range of chemicals including aucubin (an iridoid glycoside), flavonoids, tannins, essential oils and bitter phenol-carboxylic acids. The tannins are astringent; the essential oils and phenol-carboxylic acids have antibacterial properties. An infusion made from the flowering tops of *Euphrasia* is still used in herbal remedies, especially for the treatment of styes and inflammation of the eyes, and to relieve the symptoms of conjunctivitis. As an infusion taken internally, eyebright is used to decrease the nasal secretions of those suffering from hay fever or allergic rhinitis. *Euphrasia* is used by homoeopathic practitioners to treat eye disorders. Along with coltsfoot (see page 22) and melilot (see page 89), eyebright has also been included in herbal tobacco.

Given the plethora of 17th-century eulogies to the worthiness of *Euphrasia* and its supposed effectiveness in 'clearing the sight', it is apposite that in the early 1650s John Milton lost the sight of his left eye. The sight in his other eye had failed by 1667 when he published his epic, *Paradise Lost*. In book XI Milton describes:

> *Michael from Adam's eyes the Film remov'd*
> *Which that false fruit that promis'd clearer sight*
> *Had bred; then purged with Euphrasie and Rue*
> *The visual nerve, for he had much to see;*

He also included three drops from the Well of Life, so that Adam could see death and the miserable future of mankind!

Eyebright   (a) Typical flower colour   (top) Leaf   (lower) Side view   (b) Lilac flower

Fly Agaric

Amanita muscaria

Helena, the wife of Dr William Withering, initiated his interest in botany and his pioneering work on the foxglove, as told on page 52. Behind every great man .... The woman behind the American Gordon Wasson was a Russian emigré called Tina whom he married in 1927. During their honeymoon she collected some edible fungi, but her husband refused to share the fungal feast and feared he would awake a widower. Tina survived and her love of fungi, imparted in her Russian homeland, soon won over her more conservative husband. During the 1950s Wasson regularly discussed his fascination for fungi with his friend Aldous Huxley. Huxley's novel, *Brave New World* written in the 1930s had featured the non-fictional mysterious drug soma. Wasson believed that soma might have a fungal origin. He and Tina spent many years of travel and research, culminating in the 1968 classic *Soma: Divine Mushroom of Immortality*. My copy of the 1971 hardback edition, serendipitously inscribed 'ex libris Anthony Huxley', came complete with dust jacket depicting two fly agarics on the front.

If Wasson's book was to upset a few academics who were adamant that soma 'the food of the gods' was of plant rather than fungal origin, this was nothing to the uproar two years later with the publication of John Allegro's *The Sacred Mushroom and the Cross*. Based on research of the Dead Sea Scrolls and detailed knowledge of ancient Sumerian and Semitic languages, Allegro argued that both Judaism and Christianity are no more than hangovers from an ancient fertility cult centred on the fly agaric toadstool. At about the same time the red and white fly agaric was shown to have a significant role in the pagan story behind the Father Christmas image, pre-dating the Christian story of Saint Nicholas. Could one fungus species really be that important?

Fly agaric is common throughout Britain (as well as across Europe, North America, New Zealand, South Africa and parts of Asia), usually on poor, acid soil where the tree species with which it forms mycorrhizal (fungus/root) associations are found. In Britain these include birch, pine and occasionally spruce. *Amanita muscaria* typically fruits from late summer until the early winter frosts. The young egg-like stage is totally enclosed in a white veil. As the stem lengthens the fruitbody pushes through the veil, the remnants of which are left as a ring of scales (the volva) round the swollen stem base and as small fragments on the cap surface. The underside of the cap is initially encased in a second white veil, but as the fungus enlarges it parts company with the cap margin while remaining attached to the stem as a skirt-like ring.

The initially hemispherical cap expands and flattens to a diameter of up to 20cm. The veil remains take the form of concentric rings of white warts on the vermilion-coloured, marginally grooved, shiny cap. Older caps have a more matt orange or even yellow hue and the 'spots' may be washed off in the rain. The gills on the underside are pure white. One respected book on fungal identification describes the taste of the flesh as 'somewhat nutty' but most British books inform us that the fungus is fatally poisonous. This piece of misinformation has helped to feed

Two emerging fruitbodies of fly agaric

the British mycophobic culture, exemplified by the 'kick it but don't touch it' attitude to toadstools.

Amanita muscaria is rarely fatal, but ingestion (certainly not mere handling) can produce digestive upsets and other alarming symptoms worthy of its being given a poisonous label. Talking of digestion, a close relative of fly agaric known as Caesar's mushroom (*Amanita caesarea*) is edible and excellent to eat. It differs from fly agaric in its spot-free, bright orange-red cap, yellow gills and a sac-like volva. Sadly not found wild in Britain, maybe it will gain a foothold as a result of climate change, it was a favourite of the Ancient Romans, who stored it in special earthenware vessels.

Muscarine, named after fly agaric's scientific name, was first extracted from the fungus in 1869. High doses of muscarine, as found in a number of other fungi, are known to cause convulsions and can prove fatal, but fly agaric only contains a tiny amount of muscarine. Despite this, muscarine toxicity has frequently been quoted as being associated with the dangers of eating *Amanita muscaria*. In the 1950s chemists isolated bufotenine from *Amanita muscaria*. This famous hallucinogen occurs in the skin glands of toads and the discovery gave the term toadstool a whole new meaning. However, the chemists had made an error: there is no bufotenine in fly agaric, although it has been confirmed in related species of *Amanita*.

In 1967 the active chemical in fly agaric was revealed to be ibotenic acid, which on drying was converted to muscimol. Both ibotenic acid and muscimol are chemically similar to certain human neurotransmitters and produce psychoactive effects. Muscimol is up to ten times stronger than ibotenic acid, thus explaining why the dried mushroom is far more potent. About one-third

An immaculate fly agaric

of the muscimol and nearly all the ibotenic acid is passed unchanged by the kidneys into the urine within two hours of ingestion. Symptoms brought on by the ingestion of fly agaric, which may last for up to 12 hours, include nausea and vomiting, muscle twitching and lack of co-ordination followed by behaviour that alternates between lethargy and hyperkinetic activity. People frequently fall into a deep sleep from which they are not easily roused. Hallucinations or illusions are common, especially a perception that surrounding objects appear enlarged, a condition known as macropsia. Denis Benjamin of the Children's Hospital in Seattle, writing about the effects of *Amanita muscaria* in 1995, concluded 'These mushrooms are not lethal, despite their reputation'.

In the 13th century Albert Magnus wrote 'It is called fungus of flies by reason that, when crushed up in milk, it kills flies'. In the mid-18th century Linnaeus described how the pulped fungus was used in attempts to control bedbugs. The link with flies is probably more deep rooted and associated with the fly of madness—'a fly in your ear' symptomatic of its hallucinogenic properties. Of much greater interest is the long history of the ritual ingestion of dried fly agaric in parts of Lapland and Siberia. Mordecai Cooke relates some of the story in his 1862 publication *Plain and Easy Account of the British Fungi*, which mentions that intoxication could be achieved by drinking the urine of someone who had eaten the mushrooms. The clinical evidence of the excretion of the hallucinogenic chemicals outlined above reveals how 'getting pissed' would have been just as effective. Cooke also notes the macropsia suffered by those under the influence of the fungus.

It is probable that Charles Dodgson learnt of the macropsia effect from Cooke's writings and used the detail to good effect in *Alice's Adventures in Wonderland*, first published in 1865. Robert Graves (*The Greek Myths*, 1960 edition) postulated that 'Ambrosia', the wild autumn feast of the Dionysus-worshipping Centaurs and later referred to as the food of the gods, was probably none other than *Amanita muscaria*. The shamanistic use of fly agarics by reindeer-herding tribes in Lapland has given rise to the suggestion that the 'flying' reindeer of Father Christmas and his red and white outfit have their origin here. For more on this story see Patrick Harding's *The Magic of Christmas* (2004).

A few years ago my daughter Bryony transferred from Brownies to Guides and I was invited to witness her 'flying up to guides'. Imagine my astonishment when the initiation ceremony involved Bryony having to jump over a large plastic fly agaric before joining the goblin six. I am still following this up, but have made the connection with Baden Powell's early scout camps, which were held on Brownsea Island in Poole harbour. Visits to the island confirmed my suspicions; the poor acid soil supports a tree flora including birch and pine together with lots of fly agarics! I also wondered about Noddy and his fly agaric house until I read that Enid Blyton had spent many holidays on Brownsea. It appears that *Amanita muscaria* gets everywhere.

A Christmas Gift from Siberia

(For Tony and Patrick)

Santa lays his hand
on my head, and yours
a shaman
with gifts of healing
from worlds
the other side of the mushroom.

Rudolph carried him,
his nose as red as the scarlet fungus
with white spots
he shared with his master
to fly to the lands of wisdom.

Come into my house
where we huddle for warmth underground;
there's no need to knock
at my open door in the roof
where the smoke drifts out
from the fire.

Welcome Santa
brave traveller,
healer.

Gillie Bolton

Stages in the development of fly agaric fruitbodies

Foxglove

Digitalis purpurea

This beautiful flower occurs throughout the British Isles as both a wild plant and a garden favourite. In former times there were at least 100 different local names for the plant, including clothes pegs, deadmen's thimbles, fairy gloves, fingers, granny bonnets, pop-guns, long purples and wild mercury. Many books have assumed that the plant's supposedly magical association with fairies, together with its flowers (shaped like finger-coverings) has resulted in the name foxglove as a derivation of the earlier fairy gloves. However, the Old English *foxes glova* precedes any fairy glove name, despite subsequent tales of fairies providing foxes with the flowers to help them creep close to farm buildings in magical silence. It is possible that the frequency of foxgloves growing in the disturbed soil associated with rabbits' and foxes' earths may have resulted in the animal name tag. As early as 1548 William Turner recorded that the plant grew around rabbit holes.

Other sources point out that the glove may be a derivation of the Anglo-Saxon *gliew*, a musical instrument with many small bells. The references to dead men and mercury pay homage to the very poisonous nature of the plant, while the popping came from the noisy game associated with trapping air in the flower before a gentle squeeze! Shakespeare mentions long purples in one of his plays, but this was also a name for one of our common orchids and scholars argue over which plant was being referred to. While the specific scientific name aptly describes the flower colour the genus name *Digitalis* comes via the 16th century Bavarian naturalist Leonhart Fuchs, who created it from the German *fingerhut*, a thimble, and the Latin *digitus*, a finger.

Foxglove reproduces largely from seed (an average plant produces over 70,000) and the resulting rosettes of radical leaves, arising from ground level, stay green over winter before producing a tall inflorescence during the plant's second summer. Most books describe foxglove as a monocarpic biennial; that is, one that dies after it has flowered, in its second year. Grime et al. (1990) point out that foxglove is more often a short-lived perennial that flowers more than once, a feature backed up by my own observations of garden foxglove plants over the past ten years. The seeds need bare soil for successful establishment, but remain dormant for many years if buried under other vegetation, explaining the sudden flush of plants after tree felling, fire or the construction of new roads. Foxgloves are mostly associated with lightly shaded habitats on disturbed acid soil such as river banks, hedge or wall banks, recent plantations and around the roots of windblown trees.

As a flowering plant foxglove is instantly recognisable but young plants have been mistaken, sometimes with tragic consequences, for other species including mullein (*Verbascum thapsus*), comfrey (*Symphytum* spp.) and burdock (*Arctium* spp.). Foxglove leaves are toothed, oval-lanceolate and up to 25cm long, and their deeply indented main veins produce a wrinkled

Foxgloves growing in a woodland clearing

appearance. The blade narrows as winged outgrowths on the leaf stalk. Dark green above, the lower surface is more grey-green and downy in young leaves. In contrast mullein leaves are covered with dense woolly hairs on both sides, comfrey leaves are covered with bristly hairs and the broader leaves of burdock are mostly hairless. The typically unbranched, tall (to 1.5 metres) stem is usually green but some populations, for example on the Isles of Scilly, have purple stems. The stem terminates in a one-sided raceme of 20–80 nodding flowers. The pink-purple petals (the white form is more common in garden populations) fuse into a long tubular, bell shape at the apex of which are five lobes, the largest on the bottom of the flower. Inside the tube the lower surface is lined with long hairs and patterned by a series of red-purple blotches, each with a white margin.

Earlier explanations for the blotches included the belief that they were the marks left by the fingers of elves, but today they are termed 'honey guides' and facilitate the speedy visit by insects in search of nectar. Anyone who has watched bumblebees visiting foxgloves will know that they habitually start at the lowest available flower and then work upwards. Foxglove flowers are protandrous, meaning that in each flower the anthers ripen and shed their pollen before the stigma becomes receptive to pollen. The first flowers to open are those at the bottom of the inflorescence, so by the time a number of flowers are open those at the base will have shed their pollen and the stigmas will be receptive to pollen carried from previous plants visited by the bumblebee. As the insect works its way up the plant it will encounter younger flowers, producing pollen. This will be carried to another plant and as long as there is some overlap in flowering time among adjacent plants the foraging behaviour of the bumblebee, coupled with the protandry, ensures that foxgloves are cross-pollinated.

Folklore surrounding the poisonous foxglove is plentiful and includes the belief that the juice of the plant was powerful enough to get rid of any changeling child left by either fairies or witches. In Wales the juice was used to mark crosses on stone floors and these were said to have an apotropaic function in keeping witches at bay. The use of foxglove in folk medicine has had a more long-lasting impact on present-day society. Foxglove, despite its poisonous nature, has been used as a purgative, to induce vomiting and in a vain attempt to cure tuberculosis, or the king's evil as it was known. Fuchs, who put the plant in the *Digitalis* genus in 1542, commented that the plant might be useful for dropsy. Through the following centuries there are frequent references to the herbal use of foxglove in the treatment of dropsy, today known as oedema or fluid retention. Sadly the results were twofold: some were cured, others were killed by the foxglove concoction.

Medicinal practice in Georgian Britain fell into two camps, the qualified practitioner for the rich and the herbalist for the poor. Although the former prescribed drugs mostly based on plants they were still very much of the blood-letting, purging and blistering school, with an emphasis on theory rather than observation and evaluation of the drugs they used. So it was fortuitous that in 1768 Helena Cooke, one of the first patients to be treated by William Withering, a young qualified practitioner based in Stafford, persuaded her doctor to bring her wild flowers for her hobby of watercolour flower painting. There were two results of this unusual doctor-patient relationship: Helena became Mrs Withering and William turned his scientific mind to the wild flowers that were being used by herbalists in his practice area.

William Withering was also one of the first people in Britain to embrace the classification system of Linnaeus and in 1776, shortly after he and his wife moved to Birmingham, he published a four-volume treatise on the wild flowers of Britain. Much of this is an identification book, but it includes some of Withering's comments on the medicinal virtues of our native flora. Of foxglove he writes, 'a dram of it if taken internally excites violent vomiting. It is certainly a very active medicine, and merits more attention than modern medicine bestows on it'. He had already developed an interest in the plant through the work of a Shropshire herbalist, 'Mrs Hutton', who used foxglove in the treatment of dropsy. Foxglove leaf tea was, however, so variable in its strength as to render the drug very unreliable.

In Birmingham, Withering began to treat large numbers of non-paying ('sick poor') patients with leaves of foxglove that he collected only from flowering plants and carefully dried. He kept records of the dose, effectiveness and adverse reactions. Notes from 163 of these early clinical trials were included in his 1785 classic *Account of the Foxglove and Some of its Medical Uses: With Practical Remarks on Dropsy and Other Diseases*. Withering realised that it was the diuretic effect of the drug that was reducing water retention and surmised that foxglove improved blood flow through the kidneys by its action on the heart. He concluded that foxglove, 'had a power over the motion of the heart to a degree not yet observed in any other medicine'.

In the 220 years since Withering's book much has been learnt as to the exact mechanism of foxglove on the heart and *Digitalis* has become firmly established in mainstream medicine. In 1871 a French chemist was the first to isolate the cardiac glycosides that the plant contains. These include digitalin and digoxin, which act on the heart muscle so helping to control abnormal heart rhythms and also increase the force of each contraction. Digoxin and the other glycosides have not been synthesised and are extracted from foxglove and the European *Digitalis lanata* (woolly foxglove), grown commercially to supply the drugs. During and just after World War II members of the Women's Institute helped to collect and dry our native foxglove to maintain supplies of the heart drug.

Foxglove has kept its place in gardens, even though it is no longer used directly as a cure for dropsy. A favourite cultivar is the white-flowered 'Albiflora' and modern hybrids produce plants with flower colours ranging through white, yellow, pink and purple. Such garden forms differ from the native plants in that the flowers are produced all round the stem and are held more horizontally.

Sunstruck Foxglove

As you bend to touch
The gypsy girl
Who waits for you in the hedge
Her loose dress falls open.

Midsummer ditch-sickness!

Flushed, freckled with earth-fever,
Swollen lips parted, her eyes closing,
A lolling armful, and so young! Hot

Among the insane spiders.
You glimpse the reptile under-speckle
Of her sunburned breasts
And your head swims. You close your eyes.

Can the foxes talk? Your head throbs.
Remember the bird's tolling echo,
The dripping fern-roots, and the butterfly touches
That woke you.

Remember your mother's
Long, dark dugs.

Her silky body a soft oven
For loaves of pollen.

Ted Hughes

Foxglove on the Isles of Scilly

Horse Chestnut

Aesculus hippocastanum

In the early months of 2005 I presented the natural history slot on a BBC2 series based on the castle and estate at Glamis, where the late Queen Mother had spent her childhood years. Close to the castle there is a huge horse chestnut with most of its heavy, spreading branches resting on the ground. It had been a favourite play area for the young Queen Mother and her siblings. The British nature of the tree was commemorated when horse chestnut was depicted on one of four 'British Tree' stamps issued by the Post Office. Although often thought of as a British species, horse chestnut is certainly not a native tree (that is, one that arrived after the last Ice Age, when Britain was still attached to the rest of Europe).

For such a common tree that was introduced only about 400 years ago there is widespread disagreement over exactly when it arrived and who was responsible for its introduction. *Aesculus hippocastanum* is native to the Balkans, although as late as Victorian times English writers assumed that it originated in northern India. Its likely first introduction to northern Europe was via the Flemish Ambassador in Constantinople, who sent seeds in 1576 to the Dutch botanist Charles de l'Ecluse (better known as Carolus Clusius) working in Vienna. It appears that seeds or seedlings reached France in the early 1600s and England in 1612 or 1615.

Some books assume that horse chestnut was a 16th-century introduction because it is mentioned in the 1597 edition of Gerard's *Herball*. Gerard, however, is not noted for his accuracy and had probably copied his description from Lyte's book of 1578. Parkinson's magnum opus *Paradisus* (A Garden of Pleasant Flowers) of 1629, covers plants found in English gardens. He comments on the leaf shape and spreading branches, describing it as like the 'ordinary chestnut' but 'from the nuts sent us from Turkey'. He adds 'of as good value for the fruit' describing them in comparison to the 'ordinary chestnut' as being 'of a little sweeter taste'. Other sources put the introduction as late as 1633, this being the date of Thomas Johnson's much improved edition of Gerard's *Herball* that mentions a horse chestnut growing in the South Lambert garden of John Tradescant.

When John Aubrey visited the same garden in the 1690s he recorded that it included 'a very fair horse-chesse-nutt tree'. The father and son team, both called John Tradescant, were responsible for many 17th-century introductions of plants from Europe and recently colonised lands in North America. Philippa Gregory, in her historical novel based on the life of the elder John Tradescant, *Earthly Joys,* 1998, mixes fact with fiction. She portrays John presenting a horse chestnut plant to his wife Elizabeth in 1607; probably close to the correct date.

Horse chestnut, with its yearly display of spectacular flowers, quickly became a favourite among land-owners, although the earliest planting date of a tree still standing (Tradescant's garden has long gone) is no earlier than 1664. Sir Christopher Wren instigated the planting of chestnuts along the mile-long drive through Bushey Park, leading up to Hampton Court Palace.

From the mid-19th century local people would parade and picnic under the spreading branches of the avenue's flowering trees on the second Sunday in May, known in the area as Chestnut Sunday.

Sweet chestnut, *Castanea sativa*, was introduced to Britain by the Romans. It provides us not only with edible fruits, but also with useful timber and wood, comparable to that of its relative the oak. When, much later, horse chestnut was introduced, it was placed in the same genus and named *Castanea equina*. It was later transferred to the genus *Aesculus*, a word derived from the Latin *esca*, meaning food. This has been a source of confusion as it is sweet chestnut that provides the well-known edible fruits. There are several explanations as to the link between *Castanea* (used for making castanets) and horses. The modern specific name *hippocastanum* is but a Latin translation of horse chestnut. The origin may lie with the early Turkish custom of feeding the ground-up fruits to horses suffering from respiratory problems. Others stress the remarkable similarity of the plant's leaf scar shape to the outline of a horseshoe, complete with seven 'nail holes', actually the vein scars. Finally, the animal name may mirror the practice of naming plants that, unlike their lookalikes, were not very useful, such as cow parsley and toadflax.

Horse chestnut has remarkably large, shiny, brown sticky buds and large palmate leaves consisting of five or seven unstalked, toothed leaflets arranged like the fingers in a hand. The flowers are produced in 'candles', erect, branched clusters of up to 100, with five in any branch, of which only one or two open at the same time. The overlapping, crinkly edged, creamy-white petals are emblazoned with a yellow blotch. The blotches turn red as the flowers age and may be a signal to pollinating insects, thus preventing abortive visits to empty flowers. Unlike many other tree species, *Aesculus* flowers profusely in most years. The shiny seeds (conkers) are encased in a

Indian horse chestnut flowers

Palmate leaves of horse chestnut

green spiny husk and ripen by early autumn. The double-flowered cultivar 'Baumanii' does not set seed and is cultivated by grafting. It is sometimes planted where the normal form would attract too much attention from small boys! The related Indian horse chestnut (*Aesculus indica*) has stalked leaflets and pinker flowers. Its black seeds are encased in spineless, bronze fruits and do not ripen until late November.

The very British game of conkers seems to pre-date the introduction of the tree. At one time hazel nuts and even snail shells were used. The term conker may come from the French conche (a snail) or the term conqueror. Mabey (1996) points out that the earliest record of the game being played with chestnuts is not until the mid-19th century, but comments that before this time most of the trees were planted on private estates. Norwich was home to a different game: knuckle bleeders, played by swiping the enlarged end of the leaf stalk at the opponent's knuckles until they bled. Coincidentally, Norwich was recently at the centre of a row when the local council wanted to chop down chestnuts close to roads to prevent accidents to children intent on dislodging the conkers.

The soft wood of horse chestnut is of little value save in the manufacture of toys and kitchen utensils, although it was formerly used in the manufacture of artificial limbs. In older trees the heavy spreading branches frequently break off, so the tree is not the safest to picnic under. The fruits are more highly valued as the seeds contain coumarin glycosides, tannins and a high percentage of saponins. The latter produce a lather with water and chestnut extract is used in 'natural' shampoos and shower gels. The fruit extracts also have a medicinal value, especially in the prevention of cardiovascular disease (although internal use not under medical supervision is dangerous). Chestnut extract is also found in creams and bath oils used for cuts and grazes, as well as to alleviate the symptoms of varicose veins, prostate enlargement and piles. The bath oil Badedas contains *Aesculus* extract and has a picture of the horse chestnut on the packet. Early television advertisements for the product even included a horse, as well as a towel-clad lady, emerging from a bathroom.

Two homoeopathic remedies are made from the plant. An extract from the fruits is used to treat piles and liver complaints, while a leaf preparation is prescribed for the early stages of whooping cough. A particular form of homoeopathic medicine instigated by Dr Edward Bach also makes use of the plant, this time from the sticky buds. The remedy is recommended for those who repeatedly make the same mistakes in life. Coincidentally to the tree at Glamis, the Queen Mother was a great advocate of homoeopathic remedies.

The seeds, despite Parkinson's comments, are far from sweet, although they are eaten by deer and were formerly crushed and fed to sheep. In World War II they were occasionally used as a bitter-tasting coffee substitute. A rather more intriguing wartime use originated in World War I, when American imports of acetone became both expensive and erratic. Acetone is used to manufacture cordite for the armament industry. In 1915 Lloyd George asked Professor Weizmann of Manchester University to investigate alternative sources of acetone. Weizmann lived up to his name by discovering that the starch from conkers could be converted to acetone. In World War II the Women's Institute (and other organisations, including the Scouts) was involved in conker collection. Rather than receive an honour for his help with the war effort, Weizmann sought help for the repatriation of his fellow Jews, to what he called 'the sacred land'. Weizmann was involved in early negotiations that resulted in the Balfour Declaration and ultimately the establishment of the state of Israel in 1948.

> *Orlando's locks are of a good colour*
> *I'faith your chestnut was ever the only colour (Rosalind)*

Shakespeare *As You Like It*

Close up of horse chestnut flowers

England. The ensuing popularity of rock gardens, as with the building of walls, played a key role in the spread of ivy-leaved toadflax around the gardens of Britain.

The successful dispersal of the plant is attributable not solely to its deliberate introduction into gardens, but also to its remarkable biology. Indeed, its power to spread quickly is hinted at by another of its common names, mother of thousands. Ivy-leaved toadflax is a hairless perennial with a delicate trailing stem (it hangs down rather than growing upwards) that can extend to over 50cm in length. It regularly sends out adventitious roots from the leaf nodes, enabling the plant to spread vegetatively. The long-stalked alternately produced leaves are palmately veined, with the three to nine (typically five) veins ending in angular or rounded lobes and varying in outline from ivy-like to kidney-shaped. The slightly fleshy, shallowly concave leaves are rarely more than 2.5cm across. The undersurfaces of the leaves, leaf stalk and main stem are frequently tinged with purple.

The tiny, long-stalked flowers arise singly from the leaf axils over a long season, typically from April to November. Each flower is tubular and two-lipped; the upper lip ending in two lobes. The lower three-lobed lip forms two swellings (the palate) that close off the flower tube. At the rear of the flower the petal tube forms a short, down-curved projection known as a spur. The petals are lilac-coloured save for the white palate, which bears a yellow insect-guide on each swelling. The flowers are visited by bees that are capable of prising open the flower tube and inserting their long tongues to reach the nectar situated at the end of the spur.

As the tiny green globular fruits ripen, the plant executes a mechanism ensuring that some of its seedlings will gain a foothold close to the mother plant. The long flower stalks that had previously lifted the flowers clear of the foliage now bend back, forcing the capsules into any gaps between the stones or bricks of walls that form its main habitat. The ridged seeds are released when the capsules dehisce.

Strangely enough, it is not easy to introduce ivy-leaved toadflax into modern gardens, which may lack both a wall and a rockery. White-flowered plants of ivy-leaved toadflax often form extensive trailing colonies such as those adorning stone walls on the Isles of Scilly. A pubescent subspecies has been naturalised for over 30 years on waste ground in Surrey. For the past five years I have admired a colony of *Cymbalaria* plants growing on a garden wall near my home in Sheffield. The slightly hairy, more erect plants have darker purple flowers that are up to 25 mm long, twice the size of those in ivy-leaved toadflax. This is the Italian toadflax, *Cymbalaria pallida*, a more recent garden introduction that is slowly becoming naturalised, especially in northern Britain.

In southern Europe the leaves of *Cymbalaria*, which have an acrid taste similar to water cress, were previously eaten as a salad that helped to alleviate scurvy in much the same way that *Cochlearia* (scurvy grass) was used in Britain. As in the related common toadflax, the leaves of *Cymbalaria* are rich in vitamin C. Perhaps this toadflax was not so useless after all.

I do not know what plant Tennyson was thinking of when he wrote the following words, but I do know a plant that fits the poem exactly:

Flower in the Crannied Wall

Flower in the crannied wall,
I pluck you out of the crannies;
I hold you here, root and all, in my hand,
Little flower—but if I could understand
What you are, root and all, all in all,
I shall know what God and man is.

Alfred Lord Tennyson

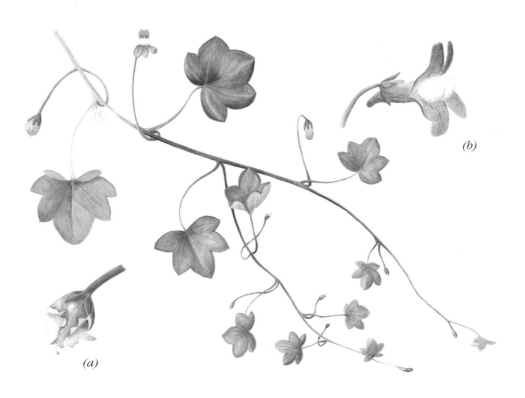

Ivy-leaved toadflax   (a) enlarged fruit   (b) enlarged flower

Juniper

Juniperus communis

When I was writing a book on British native trees I faced two problems: firstly, sorting the trees from the woody plants that are not trees, shrubs and lianas, and secondly, determining whether any trees survived the Ice Age in Britain. There is plenty of evidence to show that juniper was resident in Britain 15,000 years ago, as the Ice Age was coming to an end—but is it a tree? Scientists define a tree as a plant capable of reaching a height of at least 6 metres with a woody trunk that lacks branches from its base. Juniper refuses to be pigeon-holed; it rarely attains a height in excess of 6 metres and usually grows as a well-branched, upright shrub. Some individuals do aspire to tree status, often narrowly conical in shape, while prostrate individuals keep a very low profile, moulding to the local terrain.

Our native juniper is thus described as a very variable, erect or prostrate shrub or small tree! Many do not even give it a second glance or confuse it with a gorse bush. Grigson described juniper as growing out of sight and out of mind. Most species of tree and shrub have a habitat preference; ash on limestone, alder beside rivers, gorse on acid soils. The enigmatic juniper grows on dry chalk downs, limestone cliffs and heathland, but also in pine woods and on wet acid moorland. As a result it is locally common in southern England, the Lake District, north Wales and the Scottish Highlands. It needs bare ground for seed germination and only light grazing pressure as it becomes established. In southern England many plants date from the decrease in grazing pressure following the introduction of myxomatosis. Juniper is a declining species in Britain and little viable seed is produced. A recent survey organised by Plantlife hopes to determine the cause of this decline.

Juniper is not a true flowering plant (angiosperm) but along with our native yew (see pages 134–137) and Scots pine belongs to the gymnosperm group of plants in which the seeds are not protected by an ovary. Most gymnosperms, such as Scots pine, produce naked seeds in a cone of woody scales (hence the popular term conifers for the gymnosperm group), but in the non-conforming juniper the fleshy scales fuse to produce a berry-like fruit.

Many gymnosperms contain aromatic resins and essential oils in their leaves, fruits and trunks. These compounds are present in juniper's leaves and berries, which smell of apple or gin when crushed. They are largely responsible for the plant's medicinal, culinary and other uses. Its small evergreen, needle-like leaves are awl-shaped, tapering from a broad swollen base to a sharp-pointed apex. The grey-green upper surface sports a central white stripe. The leaves are held in whorls of three and stick out at right-angles to the twig.

As with yew (see pages 134–137), individual plants of juniper are either male or female. Male 'flowers' look like small yellow buds in the leaf axils; female plants produce tiny green buds that gradually develop into pea-size, globular green berries that take up to three years to mature, when they develop a bloomed, blue-black hue.

Juniper features widely in folklore and legend. Mary is said to have hidden the baby Jesus under a juniper tree during the flight into Egypt, giving rise to a belief that the tree could ward off evil. A juniper tree by the door to a house was deemed a useful talisman against witches, who had to count all the leaves before proceeding through the door—most lost count and gave up! It was deemed very unlucky to cut down a juniper, with predictions of the imminent death of the axeman. The Romans planted juniper bushes, the early demise of which was taken to herald the death, or assassination, of the Emperor.

The timber and foliage of juniper burns with a sweet aroma and its fumes were once believed to prevent the spread of plague. Juniper has been burnt in temples as part of religious purification rites and it was a component of the Ancient Egyptian mummification process. Juniper smoke is still used in the curing and flavouring of ham and cheeses. Juniper was also prized for its charcoal, used in the production of gunpowder. Dry distillation of the heartwood produces cade oil or juniper tar oil, still occasionally used to alleviate chronic eczema and psoriasis. Juniper's main modern claim to fame is in the use of the berries for the flavouring of gin. From the Latin *juniperus* came the French *genievre* and the Dutch *genever*, the spirit now known as gin. At one time berries were exported from Scotland to supply the gin industry in Holland.

Oil of juniper is extracted from the ripe berries that also include resins and bitters. The berries are used as a spice in the flavouring of food (for example as an accompaniment to venison or mackerel), while their constituent chemicals give them antiseptic, carminative (anti-flatulence) and diuretic properties; the resulting urine is said to smell of violets! A tincture made from the berries has proved efficacious against urinary infections such as cystitis, while juniper oil added to bath water is said to ease arthritic and muscle pain. Long-term internal use is to be avoided as juniper can cause severe kidney damage.

Juniper extract acts as a uterine stimulant and has a reputation as an abortifacient—a former name for the plant was bastard killer. It is possible that the abortifacient properties of

Juniper on the South Downs

juniper lie behind the belief, widely held in Victorian times, that excessive gin drinking would have the same effect. Less than twenty years ago Juno juniper pills were still being advertised in women's magazines under the heading 'Late? Worried?' Confusingly, the plant was also known as saffron (from the colour of the dead needles retained on the twigs) and savin, although the latter is usually reserved for the related European juniper (*Juniperus sabina*).

Juniper is commonly grown in parks and gardens. Our native species occurs as a number of cultivars including the dwarf, spindle-shaped 'Compressa' and the similarly shaped but much taller 'Hibernica'. More common are the many cultivars of Chinese juniper (*Juniperus chinensis*), originally introduced into Britain by William Kerr in 1804. These have foliage with a sour, cat-like smell that consists of juvenile shoots bearing needle-like leaves (as in *Juniperus communis*) and adult shoots bearing scale-like leaves, more similar to those of the cypresses. The mature fruits are purple-brown. Some cultivars grow to a height of at least 10 metres, but others are more suited to rockeries and troughs. Cultivars such as 'Aurea' bear a golden foliage, while those such as 'Blue Alps' exhibit a grey-blue leaf colour.

Let us now return to the last Ice Age, when the prostrate form of our native juniper, variously described as subspecies *alpina* or *nana* and bearing shorter, less prickly leaves, was a part of the tundra vegetation.

Foliage and fruit of European juniper (savin)

Extracts from *The Book of Juniper*

There is no word
and no comfort.

Only a lichened stone
is given you,

and juniper,
green juniper.

Tougher than the wind
it keeps a low profile
on rough ground.
Rugged, fecund,
with resined spines,
the gymnosperm
hugs the hillside
and wills its own survival.

...

On the brown hills
above a Roman spa
in Austro-Hungaria
the savin hides
its berries of blue wax
in a thorny crown,

...

A clear and tearful fluid,
the bittersweet genievre
is held to a wet window
above a college garden.

On the lazy shores
of a tideless sea,
the Phoenician juniper
burns a fragrant incense
in a sandy nest.

And in a Zen garden
all the miniature trees
have the perfect despair
of bound feet.

Exiled in Voronezh
the leavening priest of the Word
receives the Host on his tongue—
frost, stars, a dark berry,
and the sun is buried at midnight.

Tom Paulin

Foliage and fruit of juniper

Lesser Celandine

Ranunculus ficaria

Lesser celandine is one of the first plants to come into flower after the winter. It was formerly known as spring messenger but in recent years, as flowering dates have moved ever earlier, its golden yellow flowers have opened as early as January in some more sheltered southern sites. The much taller greater celandine (*Chelidonium majus*) is also yellow flowered but does not start blooming until April or May; it belongs to the poppy family and is not even related to *Ranunculus ficaria*.

The 1st century Roman doctor Dioscorides grouped the two species together and to him lesser celandine was known as *Chelidonium minor*. While its scientific name has changed it has kept a derivation of the earlier Latin name in the word celandine. *Chelidon* was the Greek word for a swallow and Pliny recorded the ancient belief that swallows made use of the plant's juices to sharpen the eyesight of their young. Interestingly, *Chelidonium* is still used as an eye treatment in Chinese medicine. Victorian writers link the long flowering period of greater celandine in Britain with the arrival and departure of swallows but in 2004 the first swallows, coincided with the much earlier flowering lesser celandine!

Lesser celandine is now placed in the genus *Ranunculus*, which it shares with the buttercups. Many *Ranunculus* species, including lesser celandine, grow in damp habitats—the haunt of *Rana*, the frog. The specific name *ficaria* comes from the word ficus, meaning a fig, and drawing attention to the plant's elongated fig-shaped root tubers. These underground storage organs were likened to the blisters associated with piles (haemorrhoids), a condition formerly known as 'fig'. The plant has a 'signature' (the tuber shape) relating to symptoms of the ailment that it is used to cure. In 1548 William Turner's *Names of Herbs* recorded 'Figwurt groweth under the shadowes of ashe trees'.

Figwurt, now spelled figwort (where wort means cure) was used not only as a cure for piles but also for those suffering from scrofula or 'the king's evil' (a form of tuberculosis). Here the shape of the swollen lymph glands in the neck mirrored the fig-shaped tubers. Culpeper wrote of the plant, 'with this I cured my own Daughter of the King's-Evil, broke the Sore, drew out a quarter of a pint of Corruption, cured it without any scar at all, and in one week's time'. Another plant, *Scrophularia nodosa*, has knobbly fig-shaped rhizomes and this was deemed greater figwort to distinguish it from lesser figwort, *Ranunculus ficaria*. When piles replaced fig as the common term for haemorrhoids so the plant became known as pilewort, a name it kept until the 20th century, since when lesser celandine has prevailed.

The medicinal use of plants based on signature features of the plant has largely been discredited, but not in the case of lesser celandine. As with many members of the buttercup family, *Ranunculus ficaria* contains the toxin protoanemonin but is also rich in saponins, tannins and vitamin C. The internal use of the plant is not recommended but it is still used in the preparation of herbal creams and bath preparations. These have proved efficacious, through their astringent action, in the treatment of piles and some skin disorders. Sadly, any reduction in the size of nodes (tubercles) associated with tuberculosis is not accompanied by the death of the causative bacillus, nor does the plant alleviate the more common respiratory symptoms of a disease that is once again on the increase in Britain.

Ranunculus ficaria is a plant of woodlands, hedgerows, river banks, pastures and pathsides, especially on rich, moist soil. It is a low-growing perennial with leaves and flowers that are very resistant to cold; they emerge much earlier in the year than those of other herbs. The dark green, fleshy, heart-shaped leaves have a glossy upper surface that is frequently flecked with silver or bronze. The blade margin is scalloped and the long, grooved leaf stalk has a sheathing base.

The number of flower parts is quite variable. Each shallow, cup-shaped flower typically has only three green sepals, but plants may have four or five, and one study of over 20,000 plants found half a dozen plants with six sepals. The most frequent number of narrow, burnished yellow petals is eight, but there may be as few as six and as many as thirteen. The flowers close in the evening and during dull weather. With age the petals fade to a silvery white. The open flower structure ensures that cross-pollination does not require a specialist insect visitor. There are few insects about so early in the year, but any species will do! The fruits consist of a mass of green nutlets (achenes). By mid-summer the leaves turn yellow and the above-ground parts die back. Unusually for a plant classed as a dicotyledon, the seedling has only one seed leaf (cotyledon).

Some lesser celandine plants produce little, if any, seed. They spread by division of the root tubers and from tiny bulbils. The latter are produced in the lower leaf axils after the flowers have opened. Such plants are classed as subspecies *bulbilifer*. They are more likely to grow in shady conditions and bear fewer, smaller flowers. Plants of subspecies *ficaria* have broader petals, lack bulbils and produce plenty of seed.

Lesser celandine growing as a woodland carpet

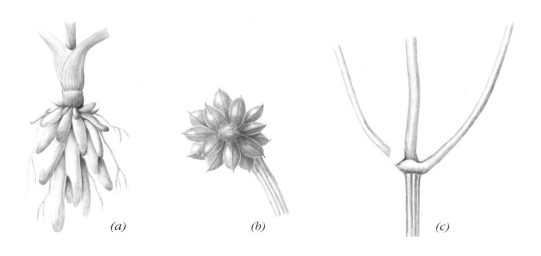

Lesser celandine   (a) Tubers   (b) Fruit   (c) Bulbils in leaf axils

Subspecies *bulbilifer* is not only more common in shady places it is also found in disturbed habitats and can become a troublesome garden weed. Despite this, several lesser celandine cultivars are sold as garden plants. One of the most popular of these is 'Double Bronze', a double-flowered form with a bronze-gold colour to the lower petal surface. Others such as 'Brazen Hussy' are grown for their chocolate-brown leaves. 'Collarette' produces very pale, double, yellow flowers with leaf-like stamen filaments.

The garden cultivation of lesser celandine is a relatively recent phenomenon, in contrast to that of *Chelidonium majus*, the greater celandine. This has such a long history as a garden plant that there is some doubt as to its status as a native wild plant in Britain. A sparsely hairy, erect branching perennial, it grows to about 90cm with large, alternate, compound grey-green leaves with lobed leaflets. The flowers occur in small upright groups. Each is protected by two hairy green sepals surrounding four, pale yellow, wrinkled petals. The flowers remain open in dull weather, unlike those of their smaller namesake.

The only point of similarity with the flower of lesser celandine, apart from having yellow petals, is in the mass of yellow stamens. The long (up to 5cm), slim fruits are not immediately reminiscent of poppy capsules, but like all members of the family they bleed a latex when damaged. In *Chelidonium* the bright orange latex is also exuded from broken stems and leaf stalks.

Early herbals indicated that greater celandine was used to cure jaundice, on account of the signature of its yellow flowers, but it has been much more effective as a diuretic and in alleviating stomach, gallbladder and other intestinal pains. Internal use of the plant is no longer recommended owing to its toxic nature, but one old remedy lives on, that of applying the orange latex to remove warts. The irritant latex can cause skin blistering and eczema and is used as a cure for these complaints in homoeopathic medicine. The highly acrid latex inhibits cell division, as does *Colchicum* (see page 84), and has been the subject of anti-cancer research.

Today greater celandine is usually looked on as a pretty, old-fashioned, clump-forming garden plant that also grows by walls and hedgerows, usually close to habitation. I always expect to see it in National Trust properties! It is sometimes represented in herbaceous borders by its double-flowered cultivar 'Flore Pleno'.

Wordsworth was more enamoured by lesser celandine and wrote no fewer than three sonnets about the plant. His poem *The Small Celandine* begins:

> *There is a Flower, the Lesser Celandine,*
> *That shrinks, like many more, from cold and rain;*
> *And, the first moment that the sun may shine,*
> *Bright as the sun itself, 'tis out again!*

In the fifth stanza Wordsworth shows his knowledge of the fact that as the flowers age so they fade and change colour:

> *The sunshine may not bless it, nor the dew;*
> *It cannot help itself in its decay;*
> *Stiff in its members, withered, changed of hue.*
> *And, in my spleen, I smiled that it was grey.*

It is fitting that there is a carving of celandine on Wordsworth's tomb in Grasmere. Unfortunately, close examination of this reveals that the carving depicts the greater celandine. Here is a classic example of the confusion caused by the use of similar English names for two unrelated plant species.

Lesser celandine flower with 8 petals

Liquorice

Glycyrrhiza glabra

In the late 1960s I was based at York University, where my doctoral research into selecting strains of grass that would grow on the colliery spoil heaps of South Yorkshire involved visits to the Pontefract area. It was only later that I discovered that coal was not the only black product to be associated with Pontefract. For the last 25 years my home has been in Sheffield where, coincidentally, a Cadbury Trebor Bassett factory produces millions of Liquorice Allsorts. Perhaps now best known for its confectionery products, liquorice has a much longer history as a medicinal herb and has played an important role in the development of a range of modern medicines.

Liquorice (also spelt licorice) has served mankind for at least four millennia. One of its earliest recorded mentions occurs on Ancient Mesopotamian clay tablets dating from about 2000 BC. The plant was also known to the Ancient Assyrians, used by the Egyptians and highly esteemed by the Ancient Greeks. Theophrastus (3rd century BC) wrote 'It is also sweet … it is useful against asthma … it has the property of quenching thirst, wherefore they say that the Scythians with this and mares'-milk cheese can go eleven to twelve days without drinking'. Pliny (1st century AD) noted that it was chewed in a powdered state by those suffering from mouth ulcers. The rations supplied to the Roman legionaries included a liquorice product.

In 1264 imported liquorice extract (at threepence per pound) is recorded in the accounts of Henry III and the plant features in the 14th-century writings of Chaucer; a character in the Miller's Tale 'cheweth lycoris to smellen swete'. *Glycyrrhiza glabra* is native to south-east Europe and western Asia. Although not wild in Britain, various 16th-century herbals mention that it was being grown in gardens. The first large-scale cultivation in England appears to have started in the late 1550s. As early as 1614 there is evidence of the Pontefract link with liquorice, or 'Spanish' as it became known in the area. The Spanish connection comes from a locally held belief that liquorice was first introduced from plant roots discovered by a local teacher on the east coast, these having been washed ashore from an Armada galleon. Other sources credit Dominican or Benedictine (Cluniac) monks as being the first to bring the crop to Pontefract when they established a monastery in the town in the 11th century.

The growing and processing of liquorice root in and around Pontefract was to bring national fame to the town in the form of the little round discs of processed liquorice extract. By the end of the 19th century demand for liquorice had outstripped supply from local growers necessitating its importation from Turkey and Spain—a more likely explanation of the 'Spanish' connection. Sadly the modern, highly automated factories now rely exclusively on imported material. Higher labour costs and a less favourable climate (compared with other areas of cultivation) sounded the death knell of liquorice cultivation in Pontefract. Before this much of the work of cultivating, harvesting and processing the liquorice was carried out by women, at a time when most of the local men worked in the collieries. Apart from the annual liquorice fair, there is little left in Pontefract as a memorial to its days as the liquorice capital of Britain.

Wild liquorice showing the zig-zag stem

Glycyrrhiza comes from the Greek *glukeros* meaning sweet and *rhiza* meaning root. There is a range of species in the genus and several have been used for medicinal purposes and in confectionery, but the plant grown in Britain has glabrous (hairless) fruits, hence the specific name *glabra*. Liquorice is a perennial herb up to 1.5 metres tall with an erect, branched green stem bearing alternate compound leaves with five to eight pairs of oblong leaflets and a similarly sized terminal one. The leaves are not unlike those of a young broad bean plant but are typically held erect. The flowers are more like those of a small pea and are produced on dense spikes from the axils of the upper leaf stalks. Flower colour varies from blue to purple and occasionally pale yellow. The smooth pods are red-brown in colour. The plants have a deep tap-root and a system of extensive horizontally creeping, woody stolons (runners), about 1–2cm in diameter.

Cultivation required deep, well-drained soil to which farmyard manure had been applied. Small portions of the stolon were planted in rows about a metre apart. For the first two years the plants remained quite small and the 'roots' (tap-root and stolons) were not harvested until the third or fourth year, when yields exceeded 4 tons to the acre. The digging up and separation of the roots was very labour intensive. Most of the Pontefract production was used to make an extract with only a little being sold as the dried root. The extract is made from crushed roots; the resulting pulp being boiled in water. The expressed liquid is then evaporated down until it thickens and can be pressed into shiny, black stick liquorice. This is combined with gelatine, flour and sugar and made into the familiar chewy 'bootlaces'.

Softer, sweeter confectionery is made by the addition of brightly coloured sugar paste as used in Liquorice Allsorts. This interesting mix of liquorice-based sweets owes its success to Charlie Thompson, who in 1899 was attempting to interest a wholesaler in Leicester in his samples of Bassett's liquorice sweets, when he dropped his individual samples. The buyer, who up to that point had not been very impressed, liked the look of the mixture and so Allsorts was born. However, liquorice lovers need to be aware that too much of a good thing can increase blood pressure (see over).

Liquorice extract is also used in the manufacture of cigarettes, pipe tobacco (it can represent 10% of the weight) and snuff, where it imparts flavour and helps to maintain the desired moisture content of the tobacco. Liquorice is also used to flavour beer, toothpaste and prescription drugs, especially if the latter contain nasty-tasting chemicals. Following the extraction process, the spent root has been put to diverse uses including the manufacture of insulating board and the preparation of a compost for mushroom cultivation. Perhaps its most surprising use is as a stabiliser in foam fire extinguishers. An extract of the spent root traps the carbon dioxide bubbles and produces a longer lasting foam.

Liquorice root as used in the pharmaceutical trade is mostly obtained from China, Russia, Turkey and Iran. It contains a number of saponin-like glycosides including glycyrrhizin, a substance first isolated in 1907. This is some 50 times sweeter than cane sugar. Other chemicals present in liquorice include bitters, tannins, starch and substances with hormonal properties. Liquorice is commonly used in cough syrups and lozenges for its expectorant action. It has also been used to treat indigestion and as a mild laxative. As a young man Spike Milligan worked as a van boy delivering to sweet shops. He later recalled that he ate so many Liquorice Allsorts that he 'had the shits for a week'.

During World War II a Dutch GP called Dr Revers noticed a diminution of symptoms among patients with peptic ulcers who took a liquorice preparation supplied by their local pharmacist. Dr Revers began to treat his patients with liquorice, but soon ran into problems with side-effects, most noticeably an increase in blood pressure and water retention. A few years later the beneficial effects of the steroid hormone cortisone (produced in the cortex of the adrenal glands) on rheumatoid arthritis were discovered. This led to the realisation that the similar anti-inflammatory and water retention properties of liquorice were caused by its constituent steroid-like chemicals. Despite being in the pea family it reduces the amount of water loss! It appears that the story of the liquorice-eating Scythians being able to go for long periods without drinking (see above) has a good basis in scientific fact.

The result of Dr Revers' work was that pharmaceutical companies began to synthesise compounds similar to those found in liquorice. One of these is carbenoxolone, which works by promoting the natural healing processes (unlike drugs that suppress gastric activity) and has been a

Close up of wild liquorice

most effective treatment for ulcers. Liquorice has also proved effective in the treatment of Addison's disease and other problems with the adrenal glands where it substitutes for a lack of the body's corticosteroids. Carbenoxolone has been described as 'a pure synthetic derivative of one particular stereoisomer of a naturally occurring plant material—the product of natural synthesis suitably modified by chemical purification and derivatization'. Synthetic liquorice is good enough for me.

One of the few remaining collieries in South Yorkshire is that of Maltby Main, not far from Pontefract. Within sight and sound of the pit is where I last came across an uncommon British native called wild liquorice. *Astragalus glycophyllos* is a hairless, sprawling, often prostrate perennial with zigzag stems. The compound leaves are similar to liquorice but the short-stalked flower heads consist of dirty greenish or cream-yellow flowers. Not having permission from the land-owner to dig up a piece of the plant I can't confirm whether it has a sweet root, but it serves as a reminder of Yorkshire's black history.

The Licorice Fields at Pontefract

In the licorice fields at Pontefract
    My love and I did meet
And many a burdened licorice bush
    Was blooming round our feet:
Red hair she had and golden skin,
Her sulky lips were shaped for sin,
Her sturdy legs were flannel-slack'd
The strongest legs in Pontefract.

The light and dangling licorice flowers
    Gave off the sweetest smells;
From various black Victorian towers
    The Sunday evening bells
Came pealing over dales and hills
And tanneries and silent mills
And lowly streets where country stops
And little shuttered corner shops.

She cast her blazing eyes on me
    And plucked a licorice leaf;
I was her captive slave and she
    My red-haired robber chief.
Oh love! For love I could not speak,
It left me winded, wilting, weak
And held in brown arms strong and bare
And wound with flaming ropes of hair.

John Betjamin

Liquorice in the herb garden at Hardwick Hall

Magic Mushroom *Psilocybe semilanceata*

When I moved to Sheffield in 1978, one of the first books I purchased was *Fungi, Man and his Environment*, a student textbook published in the preceding year. Its author, Rod Cooke, was one of a team of mycologists working at Sheffield University. The group included Tony Lyon, who became a close friend and would later join me in the writing of *How to Identify Edible Mushrooms* (1996). Rod wrote a chapter in his book called Magic mushrooms and hallucinogenic drugs. This included a short review of the ceremonial role of *Psilocybe* species in the Aztec religion of Ancient Mexico. He concluded the chapter by asking the question as to why similar practices had not arisen in other places.

> *Psilocybe semilanceata, 'liberty caps', is widespread in Europe and other north-temperate zones, and the fruit bodies contain sufficient psilocybin to have a hallucinogenic effect if they are eaten even in small quantities. This fungus is common, occurs in large numbers, is not poisonous, and is distinctive enough not to be confused with other species—yet, apparently, it has never been used even as a casual inebriant.*

Denis Benjamin in his scholarly text *Mushrooms: Poisons and Panaceas* (1995) reports magic mushrooms, 'widely used in the United States during the 1960s and 1970s … became popular in the United Kingdom and elsewhere about a decade later'. Rod's words (written in 1976) may have been prophetic, but I think that the sudden interest in the little fungi might in part be the result of a rare example of students actually reading a text book! American exchange students visiting the green field campuses of the then new universities of York, Sussex and Lancaster also helped to turn on their British contemporaries. Over the past 30 years magic mushrooms have joined the growing list of inebriants used by the British public.

As with fly agaric (see pages 46–49), it was the American banker Gordon Wasson who initiated the American student interest in species of *Psilocybe* by following the trail of earlier ethnobotanists. They had uncovered much of the history surrounding the use of *Psilocybe mexicana* and other *Psilocybe* species in religious rituals in Mexico. Wasson discovered that the mushrooms were still being collected and in 1957 he wrote an article on his own experiences with the fungi in an article for *Life* magazine entitled 'Seeking the magic mushroom'. This started a whole counterculture in the United States, with Dr Timothy Leary at its head. Later it also resulted in the most common British species, *Psilocybe semilanceata*, undergoing a change in its common name from liberty cap to magic mushroom.

In Britain *Psilocybe semilanceata* is locally common. It fruits in troops, often hundreds in a small area, from mid-summer to late autumn in acidic, unimproved sheep-grazed, pasture land, especially on higher ground and after wet weather. It also occurs on playing fields, parks and

Magic mushrooms × 2

lawns. When I moved house seven years ago and went to cut the lawn for the first time I was delighted to find that it was host to a small magic mushroom colony. Contrary to popular opinion, *Psilocybe semilanceata* does not grow on animal dung, although this habitat is home to a number of related hallucinogenic species including *Psilocybe coprophila*. The recent advent of spreading wood chippings on paths, children's play areas and flower beds has resulted in increased sightings of previously less common species that grow on wood debris. These include *Psilocybe cyanescens* and *Psilocybe squamosa*.

Those ignorant of the magic mushroom expect its size to live up to its reputation. More than one national newspaper article on the subject has appeared in recent years alongside pictures of fly agaric. Magic mushroom is much less spectacular and is included in a large group of fungi known to mycologists as LBJs (little brown jobs). Although experts may find it 'distinctive enough not to confuse it with other species' (Cooke), the same is not true of the casual collector. The tiny cap is normally less than 1.5cm deep and 1cm wide. It rarely expands, keeping a distinct conical shape like the French 'liberty cap' of its earlier name. The smooth cap is sticky and olive brown when wet but dries a pale straw colour. The cap margin is faintly striated. The young, crowded grey gills ascend to the cap apex and mature purplish black (the spore colour) with a white edge. The very slender pale straw-coloured, wavy stem is only 4–8cm high and usually turns blue at the base when picked. Two similar species often found with magic mushroom are hay cap and petticoat fungus (on or near dung); these can also induce hallucinations.

Two tryptamine derivatives, psilocybin and psilocin, are responsible for the psychotic effects of various species of *Psilocybe*. The former is converted to the latter in the human body and although colourless it rapidly turns blue when it oxidises, a feature noted at the broken end of the stem and one that assists in the field identification of magic mushrooms. The two chemicals are structurally similar to two important human neurotransmitters, dopamine and serotonin, an indication of their probable mode of action. Most of the physical effects seem to be due to stimulation of the autonomic nervous system.

The mushrooms have an obnoxious taste, best obscured with honey or chocolate (the preferred method in Central and South America). Psilocybin intoxication, which sets in within 30 minutes of ingestion, is characterised by dilated pupils, a sense of euphoria and auditory and colourful visual hallucinations that can be pleasant or frightening. Misjudgement of distances is common and time may appear to stand still. Tactile sensations are enhanced (the drug is some-times used to heighten sexual experience) and in my own experience my sense of smell, normally not very acute, was much more sensitive than usual. The effects last for about 4 hours, longer if a very high dose is taken. Occasionally the drug results in a bad trip bringing on extreme anxiety and panic attacks. Rapid heart beat (tachycardia) is one not uncommon physical symptom. The incidence of flashbacks (up to a week after ingestion) has probably been overstated.

Collins Field Guide to Mushrooms and Toadstools of Britain and Europe (Courtecuisse and Duhem 1995) notes that possession of the magic mushroom fungus (*Psilocybe semilanceata*) is illegal. This is not strictly true, but the legal status of magic mushrooms is a very grey area. The two chemicals, psilocybin and psilocin, are deemed class A drugs under the 1971 Misuse of Drugs Act, but the mushrooms themselves are not mentioned in the Act, unlike the current situation in the USA where any mushroom containing psilocybin is now a classified illegal drug.

In 2002 the British Home Office confirmed that it was not an offence to possess, sell or consume a freshly picked mushroom (that unlike cannabis occurs naturally in this country), 'provided that it has not been prepared in any way'. Dried mushrooms are frequently sold on the black market despite being deemed a preparation.

Following the 2002 'clarification' several hundred 'shroom' shops, previously restricted to places such as Amsterdam, have opened in Britain. These sell not only fresh magic mushrooms (usually cultivated and refrigerated), but also growing kits containing *Psilocybe* mycelia capable of producing the fruiting bodies. These 'shrooms' are usually species native to Mexico, Columbia, Thailand and Hawaii.

Alarmed at the way the number of such shops is mushrooming (oh dear!) the Home Office has recently changed tack again judging that 'a mushroom that has been cultivated, transported to the marketplace, packaged, weighed and labelled constitutes a product'. As I write it appears that selling the mushrooms or the kits once again constitutes a preparation—impending court cases might end up with the air of Alice and the Mad Hatter about them.

As with any drug there are problems with magic mushrooms, although unlike tobacco there is no evidence of physiological addiction. However, some people can become psychologically dependent on them and the psychotic problems are still under investigation. Bad trips are more frequent when people consume the mushrooms on their own and already have underlying worries. Very high doses and mixing mushrooms with alcohol and other drugs result in a much more powerful and potentially damaging trip. We seem to have travelled a long way in the past 40 years.

Magic mushrooms in upland pasture land

Male Fern

Dryopteris filix-mas

Although in a more primitive group than the flowering plants that make up the bulk of this book, male fern constitutes an important part of our woodland flora, was formerly used medicinally and, along with its close relatives, is still a component of many a shaded garden. In the Victorian 'fern craze' many species were collected from the wild to be grown in gardens, conservatories and living rooms. The latter were often filled with fumes and soot from gaslights and coal fires, but the accidental invention in 1829 of the Wardian case, a sealed miniature greenhouse, paved the way for a middle-class obsession.

Nathaniel Ward was a GP in London. As a keen botanist he was also interested in insects. To observe the hatching of a hawk moth, he buried a chrysalis under some moist leaf mould in a large glass jar that he covered with a metal lid. The fate of the moth is in some doubt, but what did emerge from the mould were various plants, including a fern. Ward placed the covered jar in his garden for the following three years. Moisture was recycled within the vessel so, despite no water being added, the plants flourished. By the early 1830s, encouraged by his accidental find, Ward had constructed a number of sealed glasshouses in which he was growing ferns and other plants that had not survived in his polluted London garden.

The cases not only sparked the craze of growing ferns in Victorian parlours, but also played an important role in the importation of ferns and flowering plants from abroad. Nathaniel was a good friend, and indeed a distant relative, of George Loddiges, the owner of a famous London nursery. George had already sponsored a number of plant-hunting expeditions, but the majority of plants sent by boat from more distant lands did not survive the lack of light below deck or the drying winds and salt spray on deck. In 1834, following the earlier shipment of two of Ward's cases to Sydney, Loddiges received the first live shipment of the Australian coral fern (*Gleichenia microphylla*) to reach Europe. These had survived an eight-month journey via Cape Horn in the cases lashed on deck.

Ferns differ from flowering plants in a number of ways, the most important being that they do not bear flowers, fruits or seeds. Ferns reproduce via spores born, on the undersurface of their fronds (the name given to their leaf-like parts). With the exception of hart's tongue and adder's tongue, the fronds of native British ferns are divided into sections known as pinnae. These are comparable to the leaflets of flowering plants such as ash. In many fern species the pinnae are further subdivided to give a lacy appearance similar to that found in the leaves of flowering plants such as cow parsley but these have sheathing bases and lack spores on their undersurface.

In male fern, all parts of which are poisonous, each upright frond (which can reach a length of 120cm) is typically one of several forming a loose shuttlecock-like clump. The fronds arise from a rhizome that is densely covered in brown scales. New fronds, which are mid-green above and paler below, uncurl from tight crosier shapes in the spring. The basal part of the frond is known as the petiole or stipe and this is partially covered with tissue-like, pale brown scales. The frond blade (at least 70% of the total) includes the central, stalk-like midrib (rhachis) that branches to produce 20–35 pairs of pinnae on each side. The longest of these are near the middle of the frond. The pinnae are deeply lobed and, especially near the centre of the frond, these lobes often reach to the rhachis branch; such subsections are known as pinnules. The pinnules are deeply serrated and taper to a narrow, rounded apex.

In the summer the upper pinnae on mature fronds of *Dryopteris filix-mas* bear tiny spore masses (sori) on their undersurface. These cover just over half the surface of each pinnule. Each sorus is initially protected by a kidney-shaped cover (indusium). As the spores mature from July to August the indusium falls off. Many fronds die back in the autumn but some, in well-sheltered habitats, last into the winter.

Dryopteris filix-mas is a robust fern typically growing in woodlands and its scientific name is very apt—*drys* from the Greek for a tree, *pteris* meaning fern. The specific *filix-mas* (*filix* is the Latin for fern, *mas* is as in masculine) is mirrored in the English name male fern. The name is said to have arisen to help separate this fern from the superficially similar lady fern (*Athyrium filix-femina*), which has a more delicate-looking frond in which the curving pinnules are very deeply lobed and the spore covers have a crescent-moon shape. Despite the names the relationship between the two fern species is entirely platonic.

Lady fern, which usually grows in wet shade, is not the only source of confusion with male fern. The broad buckler fern (*Dryopteris dilatata*) is common in acid woodlands where it often

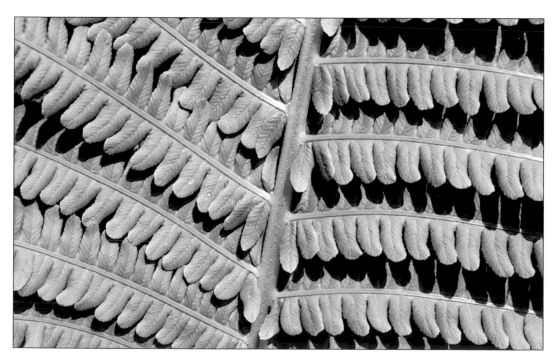

Close up of male fern

Part of a garden seat with fern design

grows alongside male fern. This species can be distinguished by its darker green, narrowly triangular frond (where the lowest pinnae are the longest) in which the pinnules are themselves divided into smaller segments and where the stipe scales are dark brown and with a translucent margin.

There are two other male ferns now considered to be separate species (previously classified as varieties or subspecies of *Dryopteris filix-mas*). The first is *Dryopteris oreades*, mountain male fern, absent from lowland Britain and most common on scree-slopes in Scotland. The other is *Dryopteris affinis*, golden-scaled male fern which, although it occurs throughout Britain, is less common than *filix-mas* especially in the south and east. The most obvious feature of this species is the presence of numerous golden, rusty brown overlapping scales on the stipe. The frond colour is usually more golden-green. The pinnules are typically parallel sided and almost square-ended, with at least two-thirds of their undersurface covered with sori. There are several distinct subspecies of golden-scaled male fern and as this is the most frequent garden plant of our native *Dryopteris* species there is also a number of cultivars. Most of these have a tight, shuttlecock arrangement of their fronds which unfurl a pale green in the spring. Cultivars such as 'Cristata' have untidy twisted frond tips.

The rhizome is the most poisonous part of male ferns, but it was previously used to prepare an extract (*filicis rhizoma*) to treat intestinal worms in both humans and domestic animals. As a vermifuge, it was recommended by Theoprastus, Dioscorides and Galen and was formerly much used to eliminate tapeworms. However, its toxic nature frequently resulted in paralysis, blindness and even death to the patient and its use by herbalists is now banned. It is still used in a homoeopathic remedy for the treatment of ulcers, septic wounds and varicose veins.

The rhizome (root in common parlance) of male fern was also burnt to provide potash for the glass-making industry (see page 15) and the ash was once used to bleach linen. The root was deemed to have aphrodisiac properties, as hinted at in this anonymous extract:

> 'Twas the maiden's matchless beauty
> That drew my heart a-nigh;
> Not the fern-root potion,
> But the glance of her blue eye.

The similarity between a bishop's crosier and the unfurling fronds of male fern was considered by those who believed in the Doctrine of Signatures (God left a sign on a plant as to its usefulness) to give the bearer of the plant a power over evil and witchcraft. The young fronds were known as lucky hands for this reason. They were also called St John's hands and were collected on midsummer's eve as a talisman against the darker days to come. Fern spores were even considered to confer invisibility to the carrier.

Fern

> Here is the fern's frond, unfurling a gesture,
> Like a conductor whose music will now be pause
> And the one note of silence
> To which the whole earth dances gravely.
>
> The mouse's ear unfurls its trust,
> The spider takes up her bequest,
> And the retina
> Reins the creation with a bridle of water.
>
> And, among them, the fern
> Dances gravely, like the plume
> Of a warrior returning, under the low hills,
>
> Into his own kingdom.

Ted Hughes

Unfurling fern croziers

Meadow Saffron *Colchicum autumnale*

This beautiful plant is both a British native and a widely grown garden flower, but its names create confusion for gardeners and botanists alike. Common names include meadow saffron, autumn crocus, naked ladies, pop-ups and snake-flower. Strictly speaking it is not classed as a crocus (part of the iris family) but is placed within the genus *Colchicum* in the lily family. The genus name comes from Colchis, in the Black Sea region of Georgia where the plant was especially abundant. The specific *autumnale* is indicative of its flowering season, but confusingly there are also autumn-flowering crocus species that are very similar in flower shape and colour to meadow saffron. Both *Colchicum* and *Crocus* species develop from an underground corm (swollen stem base). To make matters worse, saffron—the plant extensively cultivated for its stamens that provide a distinctive food dye—is a species of crocus!

Meadow saffron produces a tuft of four to five large, shiny, bright green, strap-like leaves (up to 5cm wide) in early spring. These obscure the stalked, green egg-like fruits that develop from the previous year's flowers and overwinter below ground. The leaves of crocus species are more grass-like, with a central pale midrib. Crocus leaves typically appear with or just after the flower, unlike those of *Colchicum*.

Meadow saffron's leaves wither by late summer, before the early autumn appearance of the crocus-like flowers. These consist of six graceful mauve-pink (occasionally white) tepals (sepals and petals indistinguishable) arising as two whorls of three from the end of a long thin white tube. Fifty per cent of the flower tube is below ground. The absence of leaves at flowering time is the origin of the name naked ladies.

The flowers have six stamens. Each has a long white stalk (filament) and bears an orange-yellow anther. These are each topped by the three slender white branches of the stigma which fuse at the base of the flower tube. In contrast, crocus flowers have only three stamens, the short filaments of which end in pale yellow anthers. The central much-branched, feathery stigma is a conspicuous bright orange.

Meadow saffron is a rare native in Britain and is restricted to damp meadows and woods, especially on limestone in the Cotswolds, the Mendips and along the Severn estuary. One of the earliest records is that in William Turner's *A New Herball* (1551), which mentions that the plant grew 'in the West cuntre besyde Bathe'. No crocus species are native to Britain , although several have become naturalised.

Richard Mabey (1996) comments that the flowers of meadow saffron in Wychwood, Oxfordshire, 'have the look of flowering toadstools'. In 1597 John Gerard reported in *The Herbal* on the dangers of meadow saffron: 'The roots of all the sorts of Mede Saffrons are very hurtfull to the stomacke, and being eaten they kill by choking as Mushromes do, according unto Dioscorides; wherupon some have called it Colchicum strangulatorium'. Most parts of the plant are poisonous,

*Meadow saffron—
note the six stamens*

the toxin is largely concentrated in the seeds and corm, with a smaller quantity in the leaves. The plant can prove fatal, especially to children. Grazing farm animals are also at risk , although horses and cattle typically avoid the plant, even in hay.

Symptoms of *Colchicum* poisoning have been compared to those of cholera. These include intense thirst, repeated vomiting and diarrhoea, with blood in the stools and urine. A lowering of blood pressure and body temperature may result in death from heart failure. Joint pain is another symptom resulting from meadow saffron poisoning and yet for many years tiny amounts of *Colchicum* have been used to counter the severe joint pain experienced by those suffering from gout. The plant is mentioned in the Ebers papyrus (c.1500 BC) in relation to arthritis in the big toe (probably gout).

Arabian physicians made extensive use of *Colchicum*, but it was less popular in the West. The corm shape was likened to a deformed, gouty foot as part of the Doctrine of Signatures. It was not until the late 18th century that it became more widely used in Britain and later the seeds began to replace the more unpredictable qualities of the corm. A homoeopathic remedy for gout and rheumatism is still prepared from *Colchicum*.

In the mid-1980s a student of mine, who was then in his sixties, told me a fascinating story about his treatment for joint pain. John visited his GP with the joint pain commonly known as tennis elbow, although his wrists and feet were also painful. He was prescribed the recently developed drug Feldene, marketed by Pfizer as 'A new and different anti-rheumatic'. This appeared to bring pain relief to John but was later implicated in the internal haemorrhage that he suffered. His treatment was discontinued.

Following a sudden swelling of his knees John was later confined to bed for five weeks and was admitted to hospital. After extensive tests John was given colchicine, the toxic alkaloid found in meadow saffron. John described the early effects of his treatment as being similar to a violent attack of dysentery and recalls a junior doctor cheerfully remarking 'We look on this as the drug which enables you to run before you can walk!' Within three days the pain and swelling subsided and after a week John walked out of the hospital. Subsequent attacks were quickly brought under control with colchicine. It was only

Melilot

Melilot is one of a group of related species originating from central and southern Europe. It was probably first introduced to Britain in the mid-16th century from East Friesland (now in the Netherlands) by the Tudor physic and naturalist William Turner. He placed it in the genus *Lotus*, the current home of the birds foot trefoils and concluded 'It may be called in englifth wylde lote'. Fifty years later Gerard wrote that it had become a weed in parts of Essex, where it grew in both pasture land and cereal fields. Melilot was later widely cultivated as a forage plant before being replaced by other legumes such as lucerne, sainfoin and various clovers. Today the plant is locally frequent in waste places, disturbed ground, sand dunes and railway tracks, especially in central, southern and eastern England, but is less common in Scotland.

The name melilot is associated with the plant's sweet-smelling flowers and it is a melliferous (bee-attracting) plant. *Lotos* was a term for clover and other fodder plants. The melilots not only smell of honey but also attract hive-bees along with other bees and hoverflies, all of which are able to reach the nectar by inserting their proboscises to the base of the pollen tube within the flowers. The melilots have a very long flowering period (from June to October) and produce copious amounts of both pollen and nectar. The specific *officinalis*, meaning workshop or store, is indicative of the historic importance of melilot in herbal remedies (see later). Linnaeus was the first person to use the *officinalis* tag for the plant but he placed it in the genus *Trifolium*, now the home of the true clovers. There are at least four yellow-flowered species of melilot that have become naturalised in Britain. Unfortunately in the past different naturalists have used the scientific name *officinalis* for different species! Today the name is reserved for what is now usually called ribbed (or common) melilot. This is most easily separated from its close relatives by its mid-yellow flowers and brown, transversely ribbed, hairless oval fruits.

Ribbed melilot is typically a biennial and resembles an erect, tall (to 1.5 metres), straggly hairless clover. It has a much branched, angled green stem and stalked trifoliate leaves with elongated, toothed leaflets. The small yellow, sweet-smelling pea flowers are produced in erect, slender spikes near the top of the plant and give rise to an old English name for the plant, wild laburnum. Close inspection shows that the two petals forming the keel at the front of each flower are shorter than the other three petals. The similar golden or tall melilot, *Melilotus altissima* (considered a native by some authorities), has golden-yellow flowers in which the petal lengths are all about equal, and black, net-patterned, downy fruits. Confusingly the plant is rarely taller than its relative despite both its common and scientific names! White melilot (*Melilotus albus*) is readily recognised by its white flowers and it was previously cultivated under the name of Bokhara clover.

In addition to having sweet-smelling flowers when fresh, dried melilot plants emit a strong smell similar to that of new-mown hay. For this reason melilots were formerly known as sweet

Melilot growing on a railway track

Oak Moss

Evernia prunastri

The ecological, economic and medicinal virtues of the bog mosses are extolled on pages 18–21 where the classification of the group is explained. Bog mosses are green plants that fit into the bryophyte category, a group that includes the liverworts. Oak moss is a very different kettle of fish and despite its common name is more closely related to porcelain fungus (see pages 102–105). Oak moss superficially resembles a moss, hence the common name, but it is actually a lichen (from a Greek word meaning eruption or wart), a group of organisms that modern taxonomists include in the fungal kingdom. As a very green young lecturer I remember having to give a talk about lichens to a local natural history group. At the end of the talk I asked whether anyone had any questions, at which point an elderly gentleman rose to his feet and informed me that he pronounced the word 'litchen'. End of questions!

Pronunciation of the word (and dogmatism in such areas is bound to offend) is the least of the problems where lichens are concerned. Much more taxing questions include 'what are they?' and 'which species is it?' All I can remember about lichens from my days at school was that they were presented as prime examples of the term symbiosis, loosely defined as two organisms (in this case a fungus and an alga) living together to the mutual benefit of both. For years I thought the word was sinbiosis, a seemingly more apt term for anything to do with living together without the mention of marriage; but this was before the swinging sixties.

Many lichens contain blue-green algae and since my school days these have been reclassified as cyanobacteria, a more primitive group than the algae. For this reason the above definition of lichen needs amendment. Hawksworth et al. (1995) give us: 'a lichen is a stable self-supporting association of a fungus (mycobiant) and an alga or cyanobacterium (photobiont)', but also offers: 'a lichen is an ecologically obligate, stable mutualism between an exhabitant fungal partner and an inhabitant population of extracellularly located unicellular or filamentous algal or cyanobacterial cells'. To think that some people worry about how to pronounce the word!

Forty years on from the happiest days of my life, I can turn to the wonderful writings of my Sheffield colleague, the late Oliver Gilbert. In *Lichens* (2000), his excellent New Naturalist book, he informs his readers that 'lichens are not a taxonomic group like the ferns or the liverworts; they are a fungal lifestyle …'. So there! At least the lichen fungi have their own carbohydrate factory in the form of the constituent photobiont and are therefore not dependent on an outside food source. Despite claims to the contrary, lichens growing on trees are not parasitic.

It is perhaps understandable that our predecessors confused a 'fungal lifestyle' with the mosses. Until a little over 200 years ago even scientists lumped the two together under the Latin term *muscus*, meaning moss. Today people still talk of Spanish moss (a lichen) and colleagues involved in the production of the mosaic-like pictures for the revitalised Derbyshire Well Dressing ceremonies use 'moss' from the Peak District walls, actually the lichen *Parmelia saxatilis*. Unlike

Oak moss on larch

the true mosses very few lichens are green, an unsurprising fact given that the fungal partner is very much the dominant one and very few fungi are green!

There are about 1,700 species of lichen in Great Britain and Ireland and each species, such as *Evernia prunastri*, bears the name of the fungal partner. The fungus does not occur (outside the laboratory) without its component alga/cyanobacterium. While the fungal partner is unique, particular algal/cyanobacterial species occur in a large number of different lichen species.

Oak moss is found on the trunk and branches of oak trees, but also on many other tree species, as well as on tree products such as wooden fences and gates. It is much more common in the wetter north and west of Britain and in regions not prone to high levels of atmospheric pollution by sulphur dioxide. In the 1970s the distribution of lichens, including oak moss, was mapped in a nation-wide survey. The presence or absence of certain species turned out to be a very accurate estimation of sulphur dioxide levels and showed that many rural areas were affected by pollution carried in the prevailing wind from urban and industrial sites.

Over the past 20 years many parts of Britain have experienced a considerable drop in sulphur dioxide levels and *Evernia* is once again re-invading town centres and other places that it had deserted before the introduction of smokeless zones and factory emission controls. Species of trees including willow, sycamore and ash have a less acidic bark than those such as oak, birch and alder. The acidic bark of the latter group exacerbates the effect of sulphur dioxide (which produces sulphuric acid in the presence of water). It is comforting to see that the much maligned

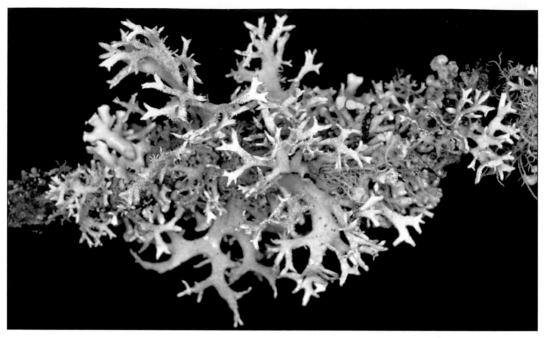

Close up of oak moss

sycamore (see pages 110–113) is among the first to be colonised by lichens (along with highly alkaline asbestos roofs) in areas previously described as lichen deserts.

The body of a lichen is known as the thallus and the majority of British lichens can be placed in one of three groups depending on their growth habit. Crustose lichens develop in the form of a crust, firmly attached to the surface of tree bark, rocks or other substrates. Foliose lichens have a thallus with leaf or scale-like sections attached to the substrate by root-like threads. The fruticose lichens, to which *Evernia prunastri* belongs, are rather bush-like and more similar to mosses. The typically much-branched thallus of fruticose lichens can be either erect or pendulous. Those that are pendulous include the aptly named beard lichens (species of *Usnea*), which festoon tree branches in unpolluted regions as well as growing on rocks, walls and cliffs near the sea.

Evernia prunastri consists of a repeatedly forking, flattened, stag's horn-like thallus which in dry/moderately polluted areas is most likely to form a small, loose, erect tuft near the base of tree trunks. In more moist, less polluted regions it grows as pendulous cushions up to 8cm in diameter, on the trunk and smaller branches of trees but also from wooden gates, fences and benches. The upper matt surface is a light greenish-grey and is covered with a network of paler wrinkles. The under surface is white. Powdery soredia (small masses of fungal threads and a few phytobiont cells) are produced at the margins and on the wrinkles; on dispersal these can form new thalli. Sexual spores are formed in disc-like apothecia, but these are rarely formed on *Evernia* except in the least polluted habitats. The lichen generally relies on asexual reproduction via its soredia.

The powdery dust from lichens such as *Evernia*, and more especially some of the crustose species, is a frequent cause of skin allergies among tree surgeons, foresters and sawmill operators. The likely cause of the allergy has been traced to chemicals known as lichen acids that are only found in lichens. These are responsible for the bright colours of some lichen species and the wonderful lichen dyes made famous by Harris tweed. The bitter taste imparted by the acids is a deterrent to slugs and other potential lichen grazers. They also have antibiotic properties and high concentrations of the acids may have evolved to prevent attack by fungi and bacteria. Usnic acid, extracted from *Evernia* and *Usnea* species, has been used as a broad-based antibiotic against infectious skin disorders including impetigo and athlete's foot. But for its low solubility in water it might have found more widespread use, although my bathroom is home to a lichen antidandruff shampoo, also based on usnic acid.

Oak moss is involved in a far more lucrative business than that associated with fungal infections and hair care. Oak moss (known in the trade as mousse de chêne), together with a similar-looking lichen called tree moss (*Pseudevernia furfuracea*—easily distinguished by the black undersurface to its thallus branches), is collected in huge quantity to supply the perfume industry. In parts of Europe, including Italy and France, and in northern Africa (Morocco) the annual collection is close to 10,000 tonnes.

The lichens are processed to form an extract known as 'concrete' or oak moss absolute. Although this has a strangely mossy smell its main purpose is to act as a fixative in perfumes and toilet soaps, ensuring that the very expensive added scents are slowly released over a long period. The lichen extract amounts to 1–12% of the finished perfume. Conservationists worry about over collection of slow-growing lichens, but the industry is more concerned that rising standards of living in the areas where lichens are collected will force up the meagre wages currently paid for the work. The next time Kate Moss is photographed in the company of other A-list celebrities remember that part of her allure is dependent on another moss that is really a lichen!

Crustose lichens growing on slate

Oxford Ragwort

Senecio squalidus

Oxford ragwort, in common with ivy-leaved toadflax (see page 58) and pineapple weed (see page 98), is an introduced species that has flourished in British habitats not previously overrun by native species. *Senecio squalidus* occurs in south-eastern Europe, including Greece and Italy, where it grows on dry rocky outcrops. It was introduced into Britain's first Physic Garden, at Oxford (now the Oxford University Botanic Garden), in the latter half of the 17th century. Jacob Bobart, the Garden's first curator, died in 1679 and was succeeded by his son, also called Jacob, who was at the helm when the plant was introduced in about 1690. The collection site is reputed to have been in Sicily, where the plant can still be found on the slopes of Mount Etna.

Senecio squalidus kept a low profile in Britain for its first 100 years but by 1794 it was noticed growing on the walls of the Physic Garden and some of the Oxford colleges. There is little evidence that the plant spread far from Oxford during the next 50 years, but in 1844 the Great Western Railway opened, linking Oxford with London and nearby large towns. The tracks were laid on a mixture of granite chips and clinker, a most inhospitable habitat for plants but not dissimilar to the natural habitat of Oxford ragwort. The initial spread from Oxford was far from rapid and it was not until the last two decades of the 19th century that it was recorded from Swindon (1890) and also Bideford in Devon (although this may result from a separate introduction).

In the 1950s D.H. Kent monitored the spread of *Senecio squalidus* from Oxford. It reached London and Wales by the early 20th century and most of its early sightings were in towns linked by the railway. The plant continued to spread in England and Wales through the 1920s and 30s. The London blitz of World War II created large areas of rubble and an ensuing population explosion of Oxford ragwort. Later the slum clearance programmes in many northern towns and cities during the 1950s and 60s created sites that were rapidly colonised by *Senecio*. Even as late as 1966 McClintock noted that it was still rare in Sussex and Ireland, and most of Scotland.

In 1970 I moved to the then New University of Ulster in Coleraine, some 50 miles north-west of Belfast. Oxford ragwort had not been recorded in the area so I was pleased to find it in the early 1970s, very close to the railway station—it was obviously still travelling by rail. It is now evident not only that the tracks have provided a habitat for *Senecio*, but also that trains have been its main method of long-distance dispersal around the British Isles. Like many other members of the daisy family the seeds of Oxford ragwort are wind dispersed and the slipstream of passing trains enhances this process. A retired train driver from the age of steam told me that the filters on locomotive air intakes regularly became blocked with wind-borne seeds in the summer. It was his job to empty the filters while the train was standing at the station—another way in which *Senecio* hitched a ride. Even in an age where tracks are regularly sprayed with weed killers, *Senecio* is still a significant railway plant.

Senecio takes its name from the Latin *senex*, meaning an old man. This is appropriate as after the white hair-like mass of seeds has blown away the remaining surface looks like a bald scalp. The specific name *squalidus* does not refer to Oxford or even the plant, but to the surroundings where it flourishes. Apart from railway tracks it frequents bare ground, brick rubble, walls and pavements, most frequently in squalid urban settings. It is a plant of lowland, urban sites (with warmer winter temperatures) and this is reflected in a paucity of records from the Lake District, parts of Wales and much of Scotland.

Oxford ragwort is a fast-growing, erect (to 50cm), bushy over-wintering annual or short-lived perennial (when its stems become woody below). The plant is typically hairless with glossy, deeply lobed leaves, the margins of which are irregularly toothed. Only the lower leaves are stalked and leaf shape is very variable. The plant has a very long flowering season, often lasting for at least nine months. In city centres it is not unusual to see it in flower in the mid-winter months. The branches terminate in yellow daisy-like flower heads (2cm across) containing central tubular disc florets surrounded by about 13 petal-like ray florets. Beneath these is a tight cup of green bracts that are black tipped. The ray florets open first and do not contain stamens, whereas in the disc florets the anthers fuse into a tube from which pollen is pushed out by the emerging stigma.

Unlike some related species *Senecio squalidus* is regularly cross-pollinated. Each hairy seed is topped by a pappus providing a parachute of hairs. An average plant can produce 10,000 seeds a year and germination takes place throughout the year.

There are various related native *Senecio* species in Britain. Common ragwort, also known as ragweed, staggerwort and mare's fart, is *Senecio jacobaea* and was formerly believed to cure horses of the staggers, a disease of the nervous system. This is one possible connection (*herba sancti Jacobi*) with St James, the patron saint of horses. The plant was also linked with the Jacobite attempts to regain the British throne for the descendants of King James, a cause finally crushed when Charles Stuart was defeated by William Duke of Cumberland at the Battle of Culloden in 1746. For many years after Culloden, *Senecio jacobaea* was known to the Scots under the name of stinking billy.

Not only does common ragwort have an obnoxious smell (like a mare's fart?), it is also highly toxic. The plant is responsible for some 50% of cases of stock poisoning in Britain and causes acute liver failure in cattle and horses when eaten fresh or in hay. Land-owners can be fined for harbouring what is a 'notifiable weed'. The caterpillars of the

Senecio sp. *growing beside a disused railway line*

Oxford ragwort growing on a wall

red and black cinnebar moth (*Tyria jacobaeae*) are immune to the poison and the orange and black-striped larvae often strip plants of their leaves.

Common ragwort is a tall (up to 1 metre) erect, biennial or short-lived perennial of poor pastureland, dunes, roadside verges and waste places. The furrowed stem typically branches near its apex. At its base is a rosette of dark green, lobed leaves. Stem leaves are more finely divided and may be cottony below. The flower heads are comparable to, but less showy than those of Oxford ragwort and are arranged in flat-topped clusters.

Since the arrival of Oxford ragwort in Britain there have been some interesting changes in groundsel (*Senecio vulgaris*), the smallest of its native relatives. This straggly, much-branched annual rarely tops 30cm. Its lobed leaves are not unlike those of the ragworts but its tiny flower heads (4 mm) usually lack ray florets thus rendering the plant much less noticeable. Groundsel is very fast growing, flowers throughout the year and is largely self-pollinated. It is a weed of cultivated ground such as gardens and rough ground including hard core (the common name is a derivation of an Anglo-Saxon word *grondeswyle,* meaning ground glutton). It frequents some of the sites such as pavements which Oxford ragwort has more recently invaded.

Occasionally groundsel is found with an unbranched stem, pubescent, less deeply lobed leaves and, most noticeably, flower heads with seven to eleven narrow, yellow ray florets. This Mediterranean form (subspecies *denticulatus*) has long been recorded from the west coast of England and Wales. As an undergraduate at Bangor I first found this unusual groundsel in the 1960s. Since then, plants much more like the typical groundsel, but with ray florets, have become increasingly common in urban habitats away from the west coast. Several such plants (var.

hibernicus) now grace the pavement at the end of my road in Sheffield. Richard Abbott, one of my undergraduate colleagues at Bangor, has investigated var. *hibernicus* and found strong evidence that it has resulted from hybridisation between our native groundsel and the introduced Oxford ragwort.

As long ago as 1948 an unusual tall groundsel with more acutely angled leaf lobes and much broader ray florets that curl back soon after the flowers open was discovered in north Wales. Genetic study has shown that this too probably arose from a cross between groundsel and Oxford ragwort, but was followed by a chromosome doubling that has prevented the hybrid from back-crossing to either of its parent species. As such it has now been designated a separate species. Welsh groundsel (*Senecio cambrensis*) is endemic to Britain and one of the few British plants that is found nowhere else in the world. Recently, Abbott discovered a different population of groundsel × Oxford ragwort plants on waste ground in York. These plants are also genetically isolated from both parents and represent another new species, *Senecio eboracensis* (York groundsel).

The coming together of the related species, Oxford ragwort and groundsel, has resulted in two novel British plant species arising in the past 100 years. Welsh groundsel and York groundsel may not look exciting, but they are living proof that selection is producing new species in front of our eyes. What would Charles Darwin have given to experience these wonderful examples of evolution in action?

Close up of Senecio *flower head*

Pineapple Weed   *Matricaria discoidea*

For a plant that is so widely distributed throughout Britain it comes as a surprise to many people to learn that pineapple weed is not native and is a relatively recent introduction. It was first recorded in 1871 when it was spotted by J.G. Baker on the coast of North Wales, near Caernarfon. An Asian native, it found its way to North America and it is from there that it appears to have been accidentally introduced to our western shores, possibly as seed that hitched a ride in cotton bales imported into Liverpool and other nearby ports. A former name for the plant is Oregon weed, although there is no proof that Oregon was the port of export for *Matricaria discoidea*.

The spread of pineapple weed was rapid, and well recorded by Victorian clergymen-naturalists among others, and by the end of the 19th century it had reached Scotland, Ireland and the more distant parts of England. By World War I, it was already frequent on bare soil at the edges of arable fields, roadsides and gateways, as well as on paths and in farmyards. These are all places of compacted soils, where trampling and vehicle pressure reduce the growth of competing native perennial species and enable the annual alien to flourish.

Matricaria seeds can germinate throughout the year and those plants arising from spring-germinating seeds grow so quickly as to be themselves producing seeds within 6–7 weeks. Each flower head may contain over 400 seeds and a large plant can produce at least 6,000 seeds. Many species in the Asteraceae (daisy family) have tiny seeds that are widely dispersed by the wind, with the help of a parachute-like feathery pappus. The seeds of pineapple weed are certainly small, on average each one weighs only 0.08 mg, but they have no obvious dispersal mechanism and most fall to the ground within a few centimetres of the mother plant.

Nearly 40 years ago I attended a conference on reproductive biology in Birmingham, where I was privileged to hear an address by an old friend of my parents, the late, enigmatic Irish botanist Professor David Webb. David gave an overview of seed dispersal and revisited some of the well-known examples of plants with seeds dispersed by agents such as wind, birds and water. It was then that he asked 'what of all the others with no obvious means of seed dispersal?' At this point he quoted a quatrain that he had learnt at school:

> *The lightning-bug has wings of gold,*
> *The June-bug wings of flame;*
> *The bed-bug has no wings at all,*
> *But he gets there just the same.*

Pineapple weed certainly got there without wings! It now appears that it arrived in Britain at a particularly opportune time. During the end of the 19th and the beginning of the 20th century, its seeds were probably carried in mud stuck to boots and the tyres of tractors and cars that were then rapidly replacing the horse as a means of transport. At the time few roads were metalled. Horses may have helped in the local distribution of *Matricaria* as the seeds remain viable after passing through their guts! In the 1960s an ingenious experiment verified the importance of vehicles in the transportation of *Matricaria* seeds. A car was steam cleaned on its undersurfaces before being driven some 65 miles through the Midlands. The journey included a number of farm visits and stops in the entrances to fields. The car was later carefully hosed down and the washing water was sprinkled over a bed of sterilised compost. Over the following months no fewer than 220 seedlings of pineapple weed emerged from the compost.

Pineapple weed

Pineapple weed has continued to spread during the past 50 years, during which time its success in the verges of main roads has been aided by its apparent tolerance to the saline conditions resulting from the winter application of salt to such roads. It has remained less common on cultivated ground and so has not become a serious garden weed.

The plant is well named; its conical flower heads are the shape of pineapples and when crushed they emit a strong aroma, perhaps more akin to tinned pineapples than to the fresh fruit. Some closely related native species such as scented mayweed have white 'petals' (actually tiny flowers that look like the sun's rays and are known as ray florets) encircling the base of the conical head of tiny tubular flowers, known as disc florets. Pineapple weed is also called rayless mayweed as its flower head consists solely of disc florets, hence the *discoidea* part of its modern Latin name. *Matrix* is the Latin for womb; the plant was formerly used to treat ailments of the uterus and this gave rise to the generic name *Matricaria* and the older specific name *matricarioides*.

Pineapple weed is a hairless, low, bushy annual with dark green, slightly fleshy, much-divided leaves ending in narrow bristle-pointed segments. The short-stalked, hollow-centred conical flower heads consist of a mass of yellow-green, four-lobed disc florets subtended by a basal row of tiny green bracts, each with a distinctive white margin. Older books specify a flowering season of May to September, but in recent years flowering plants have been observed throughout the year.

A low-growing relative of pineapple weed also has a fruity smell. This is our native chamomile (*Chamaemelium nobile*, previously known as *Anthemis nobilis*), a plant that differs in being a perennial with downy hairs on its stems and leaves. It has longer stalks to its flower heads that are surrounded by white ray florets. The name of the genus *Chamaemelium* comes from two Greek words meaning apples on the ground. A fruity smell is emitted from the leaves when the plant is walked over or sat upon.

As a wild plant chamomile is now less common than it was, being most frequent near the sea in the south of England where it is largely restricted to well-grazed turf on acid soil. In the Middle Ages it was used, as today we would use an air freshener, by strewing the plants on the floor. It has also been planted in gardens on seats and as lawns, where pressure releases the sweet smell and helps to spread the plants:

Like a chamomile bed—
The more it is trodden
The more it will spread.

For those seeking to cultivate a chamomile lawn or seat, the non-flowering cultivar 'Treneague' best serves the purpose. Another cultivar is 'Flore Pleno', with button-like double flower heads made up of ray florets. This is frequently used as an edging plant.

Flower heads of pineapple weed

German chamomile

These cultivars are of little use outside the garden, but the flower heads (mostly disc florets) of the Roman chamomile (as *Chamaemelium nobile* is also known) have been used in shampoos to highlight blonde hair colouring. The flower heads of chamomile have also been used medicinally for at least 2,000 years. Confusingly, there is another similar species known as German chamomile or scented mayweed. This plant was formerly classed as *Matricaria chamomilla*, but is now called *Matricaria recutita*. Unlike the Roman chamomile and more like pineapple weed it is an annual but it is a much taller plant than either, often attaining 50cm. It has ray florets like the Roman chamomile and a similar sweet smell, and is a plant of cultivated ground and waste places. Although many of the earlier herbals stated that German chamomile was the species used medicinally (and Roman chamomile was only of use as a lawn plant) both species contain a range of similar chemicals including essential oils (chamazulene), coumarins and tannins.

In Ancient Arabia, chamomile was the principal constituent of massage oils. Chamomile tea, made by infusing the dried flower heads in boiling water, is widely used today in Britain for its mild sedative properties and as an aid to insomnia. Of the chemical constituents some have been shown to be anti-inflammatory and others good for the digestive system, the former being put to good use in products such as nappy-rash creams.

The older British herbals make no mention of pineapple weed as it was not resident before the 1870s, but it is mentioned in North American herbals as it was used by the Native Americans in ways similar to the Old World use of chamomile. I have known people who have collected the flower heads of pineapple weed from which they have made a tea to counter insomnia. The main problem is finding the plant away from polluted roadsides or paths used by dog-walkers, so self-collection is not recommended.

What is a weed?
A plant whose virtues have not yet been discovered.

Ralph Waldo Emerson

Porcelain Fungus   *Oudemansiella mucida*

Mycologists, the posh term for those who study fungi, invariably work in university botany departments. This is confusing because botany is the study of plants, and fungi are not plants. Fungi are now grouped within their own kingdom and not as part of the plant kingdom. With the exception of a few parasitic species, plants manufacture their own food using energy from sunlight. Fungi, in contrast, require an external food source and also differ from plants in that most fungi are composed of tube-like threads, not cells. The 18th-century Swedish naturalist Carl Linnaeus, who was behind the introduction of short Latin names for plants and animals, was less successful with the fungi; he lumped them together in a group headed Chaos!

The classification and naming of plants is constantly being revised as new schemes are proposed. This may result in changes to the Latin name and even the family group in which a plant is placed. The Latin name of pineapple weed (see page 98) has been changed several times over the past 100 years. In a similar manner schemes of taxonomy and classification in the fungal world are under constant review. The porcelain fungus was formerly placed in the genus *Armillaria,* a group that includes the destructive, tree-killing honey fungus. Other mycologists placed it in the genus *Collybia,* home to a number of common species known as the tough shanks. Current authors who follow traditional schemes of classification put *Oudemansiella* (named in honour of a Dutch botanist called Oudemans) in the family Tricholomataceae. Under new schemes proposed by taxonomists, porcelain fungus is placed in a new family, Dermomolataceae. No wonder the beginner gets confused!

In addition to having Latin names, all native British flowering plants have English names. Confusingly, some plants such as wild arum (see page 126) have scores of different local names. Many of the less well-known British fungi have not acquired English names. The more common (or more spectacular) species of fungi, including *Oudemansiella mucida* have, in contrast, accrued several English names. In 2003 the British Mycological Society published a list of about 1,000 fungal species and their 'recommended English names'. It is hoped that this will not only standardise the names but also help to stimulate an interest in mycology by giving English names to those fungi that had previously only possessed Latin ones.

The list recommends porcelain fungus as the English name for *Oudemansiella mucida*. Given its pearly-white, translucent, glazed appearance this is indeed an apt epithet. In *Collins Gem Mushrooms* (1996) I used the name porcelain mushroom, but the British Mycological Society now wants to restrict the term mushroom to members of the genus *Agaricus*, which includes the cultivated and field mushroom. Other, equally descriptive alternative English names for *Oudemansiella* can be found in current field guides and include porcelain agaric (agaric is a general term for any fungus with an umbrella-shaped fruit body), poached egg fungus and slimy beech cap.

Porcelain fungus in the sunshine

The name slimy beech cap is one with feeling. Touching the cap of a young specimen provokes a sensation comparable to that of picking up a slug. The now outlawed slimy beech cap name also provides information as to the habitat of the fungus. It grows in clusters on stumps, fallen branches and standing dead trunks of beech, but is more frequent on trunks and older branches of living trees, where it sprouts from areas of dead wood associated with cast branches or lightning scars. Most books indicate that porcelain fungus is widespread in Britain, wherever beech is present, but it is far from common in the Sheffield area and it took me 25 years of searching before I found my first specimen in the Peak District. National records of the fungus show that it usually fruits from August to November but, along with many other species, its season is now extending through into the winter.

The translucent cap of *Oudemansiella* is 3–10cm across, initially domed and very slimy, but later flattening, drying (still shiny) and becoming slightly wrinkled. Its colour varies from creamy-white to a pale grey-brown. The broad white gills (ageing brown at their edges) are widely separated and are interspersed with short intermediate gills towards the cap edge. The fruit bodies grow in dense clusters and the narrow stem often curves so as to position the horizontal cap away from its neighbours. Near the apex of the stem is a membranous collar (ring).

Fortunately there are very few similar-looking fungi that grow on dead beech wood. The clump-forming, similarly sized velvet shank (*Flammulina velutipes*) has a slimy yellow-brown cap and a velvety, dark brown stem base. Rooting shank, which was known as *Oudemansiella radicata* but has now been transferred to the genus *Xerula*, grows from stumps, roots and buried wood of deciduous trees and is most frequent in beech woods. It has a larger, slimy, red-brown wrinkled cap, widely separated white gills and a very long straight stem that extends like a root under the ground.

Porcelain fungus is an edible species, but only after the slimy cap covering has been removed and the fungus has been well cooked. Given these strictures, plus its small size and insubstantial flesh, it is unsurprisingly rarely eaten in Britain. While most gastronomes ignore it others, including potters and artists, have long admired its beauty. In the late 1970s I visited the family of a friend who lived on a large farm near Armagh. After a late night quaffing copious quantities of Guinness and Bushmills, I awoke the following morning to a glorious sunny day and a severe hangover. Later as I took my camera for a walk, I squinted at the sky beyond the horizontal branches of an ancient beech tree and realised that I was directly beneath a cluster of porcelain fungi glowing like sliced lemons as they reflected the sun's rays. My hangover vanished as I used all 36 exposures of my film to record the magical scene.

Fifteen years later I was asked to lead a fungal foray as part of a residential conference that was being held in the Peak District. I had previously ascertained that the conference delegates worked for AstraZeneca (previously part of ICI) and were gathered to discuss the firm's new product Azoxystrobin (trade name Amistar). The inspiration for Amistar was a group of naturally occurring fungicides called strobilurins that are produced by *Oudemansiella mucida* and *Strobilurus tenacellus* (pinecone cap) and secreted into the substrate on which the fungi grow. Knowing that we were unlikely to find *Oudemansiella* in the Peak District, I took one of the photographs from my Irish encounter.

Research scientists had determined that strobilurins prevented other fungi invading the beech wood or pine-cone, so reserving the food for *Oudemansiella* and *Strobilurus*. More significantly it was discovered that strobilurins inhibited the growth of many of the fungi that cause mildew, rust and other plant diseases. These can severely reduce the yield of crop plants such as wheat, barley, rice and grapevines. By the 1980s many established fungicides used to combat mildews and rusts were losing their effectiveness because fungi were becoming resistant to the chemicals used. Unfortunately, the usefulness of natural strobilurins was limited by their volatility and inability to restrict fungal growth under high light conditions.

AstraZeneca scientists subsequently synthesised Amistar which, like natural strobilurin, inhibits a fungus's energy supply but is more stable in sunlight and more readily transported around treated plants. Of equal importance, and unlike many earlier fungicides, Amistar is not toxic to humans, birds or insects and rapidly breaks down in the soil. Over the past decade Amistar has become one of the most successful of the new fungicides and is now in use throughout the world.

It is perhaps satisfying to see a more 'green' approach to crop protection in the form of setting a fungus to catch a fungus. Personal satisfaction came from a joke remark that I made at the end of my session to the Peak District conference: 'Why not let your marketing department know about my picture of *Oudemansiella*?' Three weeks later I sold them the picture and a copy of it now adorns every box of Amistar. So it turned out that my Irish hangover ended with a smile.

The simplest and most lumpish fungus has a peculiar interest to us, compared with a mere mass of earth, because it is so obviously organic and related to ourselves however remote. ... It is a successful poem in its kind.

Henry Thoreau

Porcelain fungi in the rain

Porcelain fungi on a beech branch

105

Shaggy Ink Cap
Coprinus comatus

Thirty years ago I was involved in a tour of schools in Northern Ireland as part of a road show to drum up applications from potential undergraduates for my employer, The New University of Ulster. Given a tight budget we did not splash out on accommodation and our overnight stops included caravan sites. On one occasion I was delighted to find a group of shaggy ink caps among the grass beside our designated caravan. As I happily filled the frying pan I became aware of the horrified looks from the surrounding caravans and was reminded of Victorian values that considered such 'toad's meat' as decidedly unfit for the table of a Christian. To be fair on the people of Ulster there would probably have been a similar response in other parts of the United Kingdom in the 1970s. Writing in 2005 I am pleased to say that considerably more of the great British public has finally realised that *Coprinus comatus* is a good edible fungus.

Shaggy ink cap is one of the largest and most easily recognised of a group of species in the genus *Coprinus*, the ink caps. They all share certain features, including the production of black spores. Some fungi such as fly agaric (see pages 46–49) and porcelain fungus (pages 102–105) have white spores; magic mushrooms (pages 74–77) have purple-black spores; other species have brown or even pink spores. The ink caps are separated from other black-spored fungi by virtue of their deliquescent caps in which the gill tissue dissolves (auto-digests) into a black inky fluid after the spores have been released.

Shaggy ink cap is also known as lawyer's wig; in America it goes by the name shaggy mane. The fungus is rarely found as a solitary specimen, it is typically one of a crowd that may number as many as 100 individuals. It is as common in urban habitats as it is in the countryside, especially following habitat disturbance and wherever there is a buried source of rotting wood or other organic material. For nearly 20 years I taught on a mushroom course every autumn at Losehill Hall in the Peak District and on most occasions the first record of the weekend was shaggy ink cap, growing in the car park. It inhabits woodland, farmland, playing fields, lawns and grass verges, and even pushes up through pavements. Shaggy ink cap has a very long fruiting season, appearing as early as April and as late as December, with the main flush in September and October.

In very young specimens of *Coprinus comatus* only the torpedo-shaped cap is visible. Later when the stem becomes visible, the parallel-sided cap is constricted at its base as if held to the stem by a rubber band. The predominantly white cap often has a brown patch at its apex. As the cap develops the surface breaks up into tiers of shaggy white or pale brown scales, giving it the appearance of a wig (*coma* is the Latin for wig). As the base of the cap pulls away from the stem, giving more of a bell shape to the fungus, the crowded white gills can be seen on the underside. At this stage the fungus varies from 8 to 25cm tall. As the gills turn pink and then black, the base of the cap starts to deliquesce and ooze inky-black droplets. Ultimately, all that is left is a tiny flat cap on top of a slender, tapering, white, hollow stalk.

The Victorian mycologist Mordecei Cooke wrote, 'When fully expanded and melting away in inky drops, it is unfit for anything except to replenish the inkstand'. Ink recipes using the fungus mention the addition of 'corrosive sublimate' or alternatively cloves, to prevent mould, and gum-arabic to bind the solution. Artists have occasionally used the ink as a wash and it was even suggested that important documents be written with ink-cap ink; forgeries would then be detected under the microscope as lacking in spores!

Cooke also commented, 'if gathered young they afford no despicable dish'. The problem with eating ink caps is that the process of deliquescence, which usually takes place over several days, accelerates to a few hours after picking. Professor Denis Benjamin summed up the problem by advising that 'the butter should be melted in the pan before one picks them'. In their beautiful book *Mushroom*, Acton and Sandler put it more succinctly: 'eat it before it starts to eat itself'. What this fungus requires is fast cooking at a high temperature. It should be fried using minimum quantities of butter or oil. Better still, cap sections can be battered and deep-fried. The salty flavour is second to none. Slow stewing of older specimens results in a dish of hot 'slugs'; far from appetising. The answer here is to put the cooked fungi in a blender and add stock to make a delicious, if slightly grey-coloured, soup.

Food for free is all very well if one starts with the correct ingredients. Any budding do-it-yourself chef needs to be aware that there is a lookalike, *Coprinus atramentarius*, which frequents similar habitats to its close relative, the shaggy ink cap. In Britain we call it common ink cap but the Americans have a more pertinent name, tippler's bane. *Coprinus atramentarius* is typically shorter than shaggy ink cap and has a broader, more bell-shaped cap. Significantly, most of the cap is grey and lacking in scales, apart from a few small ones at the apex. The cap stays intact for many days before reaching the inky stage. It is one of the most common fungi at the base of pavement trees and in gardens in addition to the haunts of *Coprinus comatus.*

Many of the older books about edible fungi encourage people to eat common ink cap but be warned—it is only suitable for teetotallers. The problem lies in an unusual chemical present in the fungus and generally known as coprine. This is an unusual cyclopropylglutamine that interferes with the breakdown of acetaldehyde in the body. Acetaldehyde is produced when

Young shaggy ink caps in the gutter

Magpie fungus in beech leaf litter

alcohol is metabolised and, in the presence of coprine, it builds up in the bloodstream and interferes with the autonomic nervous system. The resulting symptoms, which can occur as little as 30 minutes after the combined ingestion of fungus and alcohol, include sweating, vertigo, nausea (and a metallic taste in the mouth), diarrhoea, headache and rapid heart rate. Although rarely fatal, the symptoms of coprine/alcohol poisoning commonly result in hospitalisation. Fortunately the symptoms usually disappear within a few hours.

Coprine is very similar in its mode of action to the drug disulfiram, usually marketed as Antabuse. This is used in the treatment of alcoholics, the chemical ensuring that any alcohol intake will result in the very unpleasant symptoms described above. Some mushroom books assume that the consumption of tippler's bane in a meal without alcohol is safe. However research has shown that alcohol consumed up to 72 hours after ingestion of the fungus can bring on the symptoms, not to mention any alcohol consumed up to two days before the meal. There are enough good, safe edible fungi to try without running the risk with *Coprinus atramentarius*. For more information see *How to Identify Edible Mushrooms* (Harding et al., 1996).

There is a much rarer shaggy ink cap lookalike that is mostly confined to leaf litter under beech trees. I first found a single specimen of magpie fungus (*Coprinus picaceus*) some 20 years ago when leading a mushroom course in Wiltshire. I was so moved by its beauty that I cried as I showed it to the encircled group. In 2004 the same venue produced hundreds of magpies—2004 was a wonderful year for mycologists. When young the white cap of magpie fungus is easily confused with that of the shaggy ink cap, but older specimens exhibit a mottling of white patches on a grey-black background. This beautiful fungus has a rather unpleasant smell, akin to tar or coal gas, and apart from the ethics of consuming such an unusual species some authorities indicate that it can cause stomach upsets, so it is best avoided.

Mushrooms

Overnight, very
Whitely, discreetly,
Very quietly

Our toe, our noses
Take hold on the loam,
Acquire the air.

Nobody sees us,
Stops us, betrays us;
The small grains make room.

Soft fists insist on
Heaving the needles,
The leafy bedding,

Even the paving.
Our hammers, our rams,
Earless and eyeless,

Perfectly voiceless,
Widen the crannies,
Shoulder through holes. We

Diet on water,
On crumbs of shadow,
Bland-mannered, asking

Little or nothing.
So many of us!
So many of us!

We are shelves, we are
Tables, we are meek,
We are edible,
Nudgers and shovers
In spite of ourselves.
Our kind multiplies:

We shall by morning
Inherit the earth.
Our foot's in the door.

Sylvia Plath

Mature shaggy ink cap

109

Sycamore

Acer pseudoplatanus

I must declare my hand from the outset; I like sycamore, but then I have always been one who supports the underdog (a phrase borrowed from our woodland history: the underdog was the unfortunate man in the pit under the tree, holding the bottom handle of the saw). I know that I am in the minority for there are still thousands of naturalists who gain immense satisfaction from uprooting sycamore seedlings and who denounce this immigrant for usurping its elders (or other tree species) and betters; racism is alive and well established, rather like sycamore, in Britain. Evelyn, writing in the 17th century, was among the first to tarnish the reputation of sycamore, decrying its mushy leaf litter and hoping that the trees would be 'banish'd from all curious gardens and avenues'.

During the 1970s I taught plant ecology to undergraduates in Ulster. Not far from the university was a wonderful area of cliff vegetation, complete with feral goats. Among the interesting plants of the reserve was a rare filmy fern, a species that even most naturalists would mistake for a moss. As part of an exercise in mapping the detailed distribution of these ferns my students measured a range of microhabitat factors such as aspect, light intensity, temperature and evaporation rate. They also recorded the position of large rocks and mature trees. Most of the mature trees were sycamores and the management plan for the area was already proposing the removal of these non-native species, while promoting the equally non-native goat population. The student work showed that the shade and resulting higher humidity provided by the sycamores was a key factor in ensuring the continued existence of the filmy fern at the site. It was time for the planners to go back to the drawing board (probably made of sycamore).

Naturalists have dismissed sycamore as playing host to relatively few species of invertebrate and thus, by implication, not attracting many birds. A classic account by Kennedy and Southwood in the early 1980s revealed the number of invertebrate species that feed on the leaves of native and commonly introduced trees in Britain. Native trees, especially oaks and willows, dominated the top of the list, with sycamore near the bottom. This was more than enough to underline sycamore's bad name. However, as our politicians constantly remind us, figures can be produced that serve to endorse pre-conceived ideas; it depends on exactly what the figures are measuring. In the 1980s' study the numbers were of different species of invertebrate feeding on the leaves. If instead we look at the total biomass (weight of living organisms) of the aphids that inhabit sycamore leaves we find that it is very similar to that found on oak leaves.

Significantly, the heavy aphid infestation on sycamore usually occurs later in the year than that on the oaks, ensuring a continuity of food supply to fledgling birds and putting a very different light on the ornithological value of *Acer pseudoplatanus*. Current research reveals that as a habitat for a wide range of organisms, sycamore is much more important than had previously been believed. As an example, the story related on page 92 indicates that the bark of sycamore is

Sycamore silhouette

one of the first places where many lichens become re-established following a reduction in atmospheric pollution.

The value of sycamore to wildlife may be in doubt, but so too is its status. Thirty years ago when I was learning to identify trees I relied on Mitchell's *A Field Guide to the Trees of Britain and Northern Europe,* which included under sycamore 'Introduced probably by the Romans'. This early date was later revised and most tree books (my own included) now assume that it was introduced from mainland Europe in the Middle Ages, probably during the 15th or 16th century. *Love's Labour's Lost*, which Shakespeare penned towards the end of the 1580s, includes:

> *Under the cool shade of a sycamore*
> *I thought to close mine eyes some half an hour*

Parkinson, writing in 1629, notes that it was 'cherifhed in our Land onely in Orchards, oe elfewhere for fhade and walkes … and hath no other vfe with vs that I know'.

There is no indication that it was widely planted by the early part of the 17th century or had become naturalised by that time. Church carvings, notably those of the late 13th century in Christ Church Cathedral in Oxford, that are thought to be of sycamore pre-date this period, but many dismiss them as the work of imported carvers or cases of mistaken identity. As for our native trees (those species that arrived at the end of the last Ice Age, when England was attached to mainland Europe by a land bridge) field maple is on the list but sycamore is not.

How do we know which species of tree arrived in Britain as conditions warmed up from around 12,000 years ago and before the English Channel formed about 8,000 years ago? Most of the evidence comes from analysis of pollen remains in peat bog and lake deposits, plus infrequent finds such as the 'bog oaks' preserved in waterlogged conditions. In a recent article, Ted Green (2005) attempts to re-evaluate the history of sycamore in Britain. He points out that the pollen of sycamore is apparently indistinguishable from that of the field maple that is taken to be native, at least in England and Wales. In the latter the word *masarnwydd* is used for both sycamore and field maple (*Acer campestre*). The lack of early artefacts made from sycamore can be ascribed to the fact that its wood decays more readily than most of our native species.

The etymology of sycamore simply adds to the above confusion. Parkinson called it the 'great maple or sycomore tree' and a few years later Evelyn (1664) headed chapter XII of *Sylva* with the title 'Of the Sycomor'. He added that it was also falsely known as wild fig-tree. The older spelling sycomore originates from the fig-mulberry, *Ficus sycomorus* (*sycos* or *ficus*, the fig, and *morus*, the mulberry), a tree with edible fruits mentioned in the Bible. The name great maple alludes to the much larger leaves (and trunk size) than that of field maple. In Scotland sycamore is known as the plane tree (despite the fact that planes have alternate leaves, in contrast to those of the maples, including sycamore, which are produced in opposite pairs) and this confusion is alluded to in the specific name *pseudoplatanus*.

Despite being described as a weed by many, sycamore provides a very useful source of timber. Aaron and Richards (1990) in their book *British Woodland Produce* describe sycamore as 'A most useful wood with moderately good strength properties (roughly equivalent to those of oak), good stability and a pure white colour'. Brough, the editor of *Timbers for Woodwork* (1947), went further and commented, 'The sycamore yields a beautiful timber. The grain is close, even and fine … At times it is figured with a beautiful rippled mottle'. This 'fiddle-back' grain is in demand for the production of veneers and for the manufacture of the sides, backs and stocks of violins and other stringed musical instruments.

The pale-coloured, fine-grained wood does not stain or taint material with which it comes into contact and is ideal when associated with food or textiles. It is still used to make rollers, draining and chopping boards, and wooden spoons. Before plastic took over it was the wood of choice in the manufacture of bobbins and cotton reels. Its soft wood lends itself to being carved and it was formerly used to make wooden moulds to shape butter pats into figures of flowers or animals. Sycamore grows quickly after cutting back to a coppice stump and the resulting wood can be easily turned to provide chair legs, toys and games.

Sycamore trees also have a place in our folklore, especially in Scotland where some of the largest and oldest trees are to be found. Dool (grief) trees, mostly sycamores, were used by the powerful land-owners as scaffolds from which to hang transgressors, who were mostly vagrants and members of the working class. The sycamore also has a strong association with the working classes in England, with one particular sycamore being cared for by the Trades Union Congress. The tree grows in the village of Tolpuddle in Dorset, where it is known as the Martyrs' Tree. It was under this tree that six agricultural labourers had met in the 1830s to discuss the increasingly harsh treatment of their fellow workers at a time when the agricultural worker's wage was being reduced. Their leader, George Loveless, gained support from the Grand National Consolidated

Trades Union and later the Friendly Society of Agricultural Labourers was formed. In 1834 Loveless and six of his associates were arrested, not for forming a trade union (an Act in 1824 had made these legal) but because they had formed a 'sworn union' and this oath procedure made them open to a charge of mutiny. They were transported to Australia.

Some years ago I visited Birnam where the remnants of the wood that did 'come to Dunsinane', as prophesied by the witches in Shakespeare's *Macbeth*, can still be visited by the riverbank. Here there are some large venerable old oaks but beside them is a number of huge-girthed sycamores. As with most mature sycamores these display the wonderful whorled pattern in the bark and an imposing outline. Sadly, sycamores are regularly killed by the bark-stripping actions of another introduction, the grey squirrel, and by sooty bark disease. We may not yet talk of 'little helicopters to big sycamores grow', but perhaps it is time to take a fresh look at *Acer pseudoplatanus*.

Whorled bark of sycamore

Tulip Tree   *Liriodendron tulipifera*

In the latter half of the 16th century tulips (*Tulipa* spp.) were exported from Turkey to northern Europe. Carolus Clusius was one of the first to grow the newly imported bulbs when he was living in Vienna. He was sent tulip bulbs from Constantinople by Ogier de Busbeq; the same source implicated in the introduction of horse chestnut (see pages 54–57). Clusius later moved, with his bulbs, to Frankfurt and then to Leiden. The first four decades of the 17th century saw interest in the tulip develop into the passion centred on Holland where, in the mid-1630s, 'tulipomania' became the dot.com bubble of its day.

One Englishman caught up in the tulip craze was the elder John Tradescant, who was one of first to grow horse chestnut in Britain. In 1610 Tradescant, then head gardener at Hatfield House, the Hertfordshire home of Sir Robert Cecil, journeyed to France and Holland where his shopping list included tulips. He bought 800 bulbs at 'ten shillings the hundred' which, as Anna Pavord (1999) points out, was a total bill about equal to six months' wages for a gardener at the time. By 1636 John was working for himself and was one of many to buy tulip bulbs in the speculative days of the tulip boom. When, in 1637, the market crashed, John was nearly ruined and it seems likely that the need to recoup his losses was one reason behind the journey of his son, also called John, to Virginia to seek new plants for his family to sell.

New World plants introduced by the younger John Tradescant include herbs, shrubs and trees that still adorn our parks and gardens today. John made several trips to Virginia where seeds from a local tree were to help turn around the family fortune, although sadly the elder John died while his son was exploring distant lands. The American tree was *Liriodendron tulipifera*: tulip tree. In *Virgin Earth*, the best selling historical novel about the younger John Tradescant, Philippa Gregory includes a fictitious conversation between John and a prospective client: 'And we have many more rare trees that I have brought back from Virginia. I have a tulip tree with great green leaves that flowers with a blossom like a tulip as big as your head.' As we shall see this last boast is a bit of an exaggeration—either salesman's talk or poetic licence.

John Evelyn, writing in his now famous tree book *Sylva* in 1664, indicated that the tree was still something of a rarity:

> They have a poplar in Virginia of a very peculiar shap'd leaf, as if the point of it were cut off, which grows very well with the curious amongst us to a considerable stature. I conceive it was first brought over by John Tradescant, under the name of the tulip-tree (from the likeness of its flower), but is not, that I find, taken much notice of in any of our herbals: I wish we had more of them; but they are difficult to elevate at first.

Tulip tree showing leaves and flowers

The reference to poplar is an interesting one, for in 1656 John Tradescant advertised the tree as 'Tradescant's white Virginia poplar'. Today in America the tree is called the tulip poplar and timber from the tree is known as yellow poplar. The apparent similarity with poplars is in part due to its tall, slim outline and little-branched, columnar trunk when grown in its native country. In addition, the leaves of tulip tree have long slender stalks with the result that they flutter in the breeze like those of our poplars and the related aspen. Throughout the 18th and 19th centuries the tree became much more fashionable, especially in southern England where it was widely planted as an amenity tree.

Despite some common features, tulip tree is not related to the poplars but is in the same family as the magnolias, which it resembles in the form of its flowers and fruits. However, like the poplars *Liriodendron* grows best on deep, moist, rich soils. It is probable that the trembling of the leaves increases evaporation from their surfaces and explains the tree's requirement for a moist habitat. Tulip tree is slow growing as a seedling and its brittle roots mean that it does not transplant well, unless grown as a container specimen. Evelyn was correct when he said that they were 'difficult to elevate at first'. Once established, however, it is a fast grower and attains a great height, especially in the warmer climate of southern England. It grows more slowly in the north and is rarely planted north of Edinburgh. It is long lived, with one tree in Surrey now past its 325th birthday, but older trees are prone to storm damage.

Many features of this deciduous tree are unusual, including its shiny red-brown buds. These are about 1cm long and are borne on short stalks. They are flattened with a curved tip and have been described as looking like a beaver's tail. The leaves look as though they have been cut out by a small child wielding a pair of scissors. The slightly indented, truncated leaf apex is, together with the very long petiole, the most arresting feature of what is basically a four-lobed leaf. The two

115

Close up of tulip tree flower

lower lobes spread outwards, but the upper two are parallel-sided, apart from those on young plants and on sprout shoots where they have a more waist-like appearance.

The leaf outline is not unlike that of an ageing tulip flower, but the tree gets its name from the remarkable, tulip-like, fragrant flowers that are produced in June and July. These are about 7cm across (not quite head-size!), initially cup-shaped, with six oval, waxy, erect petals surrounding a mass of creamy-yellow stamens. In the centre are the carpels in the form of a green, pointed spindle. Each petal has a dark blue-green base, an orange middle and pale green-cream tip. The over-wintering fruit is like a narrow, erect, papery brown pine-cone. *Liriodendron* rarely flowers until it is at least 20 years old and then only in certain years. Frustratingly, most of the flowers are produced high in the crown and are frequently hidden among the foliage. For me the flowers are a bonus; the leaves are enough on their own, particularly as they unfold in spring. Later in the year their butter-yellow autumn colours last well into November.

In America the tree was widely planted in early colonial times as a shade plant around houses. Its soft, even-grained wood has many uses. Native Americans felled it for making large canoes, and more recently as 'whitewood' it has been used in house building, for kitchen furniture, piano parts and in the production of plywood. In Britain it was used in the early 20th century by Morgan Motors, to provide the framework for some of their early cars. The wood is occasionally streaked with bands of yellow, green or mauve, when it is sought after by cabinet-makers.

Until 1875 it was believed that the genus *Liriodendron* only contained one species, *tulipifera*. In 1901 Ernest (better known as 'Chinese') Wilson was the first person to send back to this country seeds of a second species. These were collected during Wilson's first trip to China, when he also succeeded in collecting seeds of the dove tree (see pages 30–33). Wilson later recorded in his book *A Naturalist in Western China* (1913): 'Near the summit of Hsan-lung shan, which is composed of Cambrian-Ordovician limestones, the Chinese Tulip tree (*Liriodendron chinense*) is common in the woods …'. One colloquial name for the tree is 'Wo-chang-chin' or goose-foot, a reference to the leaf shape which, although similar to that of the American species, has a more constricted (waisted) middle.

Chinese tulip tree is a much smaller tree than its American relative, which is capable of attaining a height of over 50 metres. In addition to their different shape, the leaves of *Liriodendron chinense* have a distinct blue-green underside. The smaller, similarly shaped flowers have petals that are green on the outside and green with yellow veins within. They typically flower towards the end of summer, some three weeks later than *L. tulipifera*. The tree is not common in Britain and is generally much less vigorous. In 1970 the two species were successfully hybridised, so correct identification of specimen tulip trees will become even more difficult!

Tulip Tree

Whose candles light the tulip tree?
What is this subtle alchemy,
That builds an altar in one night
And touches the green boughs with light?
Look at the shaped leaves below
And see the scissor-marks they show,
As if a tailor had cut fine
The marking of their every line!

These are no leaves of prudery
Hiding what all eyes should see;
No Adam and no Eve lie hid
Below this leafy coverlet:
The long limbs of that flower-hid girl
Would need no leaves to twist and curl,
The markings of that leaf-hid boy
Want no flowers to mar and cloy.

And so these leaves and their lights
Live only for the tulip-rites
At this altar of bright fires,
Sweet-scented, lest their ardour tires;
Leaf, and flower, and scent are all
Alive for this lit interval:
Between two winters are they born
To make great summer seem forlorn.

Sacheverell Sitwell

Leaf of Chinese tulip tree

Valerian

Valeriana officinalis

After many years of giving talks on herbal medicine to horticultural societies I can now anticipate the whispered comments that inevitably ensue when I show a slide of valerian. Just as the medicinal meadow saffron (see page 82) is frequently confused with species of autumn-flowering crocus, so the ubiquitous garden plant, wall valerian (*Centranthus ruber*) is often taken to be its near neighbour, the much more efficacious valerian.

Valerian is a frequent native British plant and although occasionally grown in cottage garden-style herbaceous borders it is much more common in the wild, where it grows in a wide range of habitats including river banks, ditches, woodland edges and dry limestone grassland. *Centranthus ruber*, by contrast, was introduced into our gardens from the Mediterranean during or before the 16th century. In 1597 Gerard noted that it grew 'plentifully in my garden, being a great ornament to the same, and not common in England'. By the 19th century it had become naturalised on walls, cliffs and other dry habitats. It has two English names, red valerian and wall valerian, the former reminding us that the flowers of this species are typically red (hence the specific name *ruber*) although some plants have white flowers; the latter name is indicative of its habit of growing on many a garden wall. It has a very long flowering period, the flowers attracting bees and other insects.

Valerian has a host of earlier names, including all-heal, cat-trail and phu. The latter is said to have originated with Galen, the famous 1st century physician; phu being his exclamation resulting from the overbearing smell of the dried roots of valerian. Time has not improved the smell, which I liken to the odour emanating from my teenage son's sweaty trainers. Chemists have shown that one of the root's constituents, valerianic acid, is also found in human sweat! It is this strange smell that makes the plant seemingly attractive to cats, which have frequently been observed digging out its roots. Knowledge of the alluring scent of valerian root was formerly put to good use by cat-catchers. Other mammals, including rats, can also be strongly attracted to the plant, hence the intriguing theory that the valerian root in his pocket was a more powerful attractant than the Pied Piper's music! The root of red valerian is considerably less aromatic.

All-heal, like the Latin name for valerian, points to its importance as a medicinal plant. *Valere* comes from the Latin meaning 'to be well' and may also relate to an early herbalist known as Valerius or the old Roman province of Valeria. *Officinalis* means 'used in medicine' and comes from a Latin word originally meaning shop, later herb store or pharmacy. *Valeriana* has a long history as a herbal cure (unlike *Centranthus*), but before investigating its claims as a cure-all we need to check on its identification.

Valerian is a rather variable perennial with normally just one erect, tall (up to 150cm) grooved, non-woody, hollow green stem that dies back to ground level in the autumn. (Red valerian is shorter, much branched and with smoother, greyer stems going woody at the base and often over-wintering complete with at least the lower leaves.) The paired compound leaves of valerian are dark green and hairless, and all but the lowest are composed of an odd number of narrow leaflets that may or may not be toothed. (Red valerian's opposite leaves are simple, often glaucous, slightly fleshy and broadly ovate in shape.)

Flowering in *Valeriana* is from June to August, whereas *Centranthus* plants flower over many months of the year. Both species produce tiny funnel-shaped flowers massed in umbel-like heads up to 10cm across. In valerian the deep pink buds open to whitish pink flowers that smell strongly of vanilla, whereas the less aromatic *Centranthus* flowers are typically red or white. The seeds of both species are wind dispersed with the help of feathery projections from the remains of the flower.

The part of the valerian plant least often seen is the basis for its importance as a medicinal plant. The plant has a short (2–4cm) vertical rhizome (underground stem) from which a mass of roots (each up to 15cm long and 2mm across) diverge like so many slender parsnips. The rhizome and roots contain an array of chemicals including a complex, yellow-brown volatile oil that includes the already mentioned unpleasantly smelling valerianic acid. In addition, there are several esters

Valerian in a herb garden

known as valepotriates, together with traces of alkaloids, tannins and bitters.

Valerian was cultivated in England for many years to provide rhizomes and roots for pharmacists. The main areas of cultivation included Hertfordshire, West Suffolk and Derbyshire, where those employed in its production were referred to as valerie growers. The Derbyshire operation was centred on the villages around Chesterfield and *Flora Britannica* (1996) includes a diary account from the 1860s explaining the nature of this cottage industry:

> *He and his companion would set out early on a spring morning walking from Clay Cross towards Chesterfield. Each of the pair has an empty bag rolled up and carried under one arm, and he also has a small fork of wrought iron. They proceed together for several miles for tho' the seedling valerian plants of which they are in quest are to be found in the woods on each side of the road they are not in sufficient abundance to justify a break in their*

journey till near Chatsworth woods. The seedlings are now developing a pair of rough leaves and where these appear in abundance the little iron forks are applied in lifting them from the leaf mould of which the soil here mainly consists. They collect sufficient seedlings for the plot of land already set apart and prepared for the replanting. This was done in regular rows and at the right distance apart to allow for growth. After which little attention or care was required beyond keeping the ground free from weeds. However vigorous the growth above ground, this is all rejected and allowed to waste, the root being now the only portion of the plant with any value …. A little field, barely three-quarters of an acre in extent, in one season grew a crop of valerian which realised seventy-five pounds. The bulk of this produce is now exported to the USA.

A booklet produced by His Majesty's Stationery Office during the early 1940s records that a few growers were still in operation at that time. It explains that the sets (either seedlings or offsets from older plants) were planted out nine inches apart in rows 18 inches apart in well-manured land. During summer the flower stalks were removed to promote greater root growth and the plants were scythed down in September or October before digging up, washing and drying the rhizomes and roots.

The great Greek scholar Dioscorides recommended valerian as a treatment for epilepsy and this use was revived in the Middle Ages. Valerian extract is known to be antispasmodic, thus corroborating its historic use in the treatment of epilepsy. It is also carminative, explaining more

Wall Valerian

modern usage that includes alleviation of the symptoms of gripe and irritable bowel syndrome. Recent research has shown that both the volatile oil and the valepotriates are sedative in their modes of action and this supports the most common usage of valerian, as a sedative or tranquilliser in the treatment of anxiety and insomnia. As such, it is still found in many herbal preparations and herb teas sold under the pretext of promoting a better night's sleep.

Valerian is also a pain reliever (some of its constituents having mildly anaesthetic properties), especially where the pain is associated with tension. Many years ago, when it was still easy to obtain an alcoholic tincture of valerian root, I experienced the pain of shingles and found relief from the pain and resulting sleepless nights by taking a preparation of valerian. My parents' generation had reason to use valerian during the nights spent in air-raid shelters (at a time when Hitler is said to have made much use of it). They were the first generation to be offered the more modern synthetic sedatives that were developed after World War II.

As with most medicines the use of valerian is not without its problems. Continued use and high dosage can result in severe headaches, palpitations and other side-effects. In the search for safer tranquillisers, synthetic sedatives were developed in the latter half of the 20th century. As such it is ironic that the drug that replaced valerian and was even given a similar name, Valium, is, along with other Valium-type drugs (the benzodiazepines) implicated in both physical and psychological addiction. Coupled with the addiction are side-effects including confusion and depression; a sad story that has implications for more than a million drug addicts in Britain.

Valerian growing wild in the Peak District

White Bryony

Bryonia dioica

Having a daughter called Bryony (using the modern spelling; the plant was formerly known as briony) was reason enough for me to include the plant in this book. Its fascinating biology and folklore, plus its use in medicine (all parts of the plant are poisonous), together with the beauty it imparts to a hedge in mid-winter, all give *Bryonia* a broad appeal.

White bryony is our only wild plant that belongs to the melon or cucumber family, other species of which are more at home in warmer climates or in the vegetable garden. Both the English name bryony and the Latin *Bryonia* come from the Greek *bruein*, meaning to grow vigorously. This is exactly what the scrambling, angular, much-branched, brittle stems do as they emerge from the plant's rootstock in the spring. By autumn, when the above-ground parts die off, the stems can have grown 4 metres in length.

Unlike other climbing hedgerow plants such as woody nightshade and honeysuckle, *Bryonia* has a non-woody stem and it gains support by means of the long, unbranched, neatly coiled tendrils that arise beside the base of the leaf stalks. Each tendril includes a section coiled in one direction followed by a relatively straight section and then a terminal part neatly coiled (often round other plants) in the other direction. This arrangement functions as a perfect shock absorber and, in drawing the plant closer to its support, minimises wind damage.

The alternate, long-stalked, palmately veined, coarsely hairy, matt green leaves are typically five-lobed and reminiscent of vine leaves. In classical times the plant was known as the white vine. The true grapevine has very different flowers (and fruits!) and its branched tendrils lack the double coils of those produced by white bryony. The attractive leaves and neighbouring tendrils have long enamoured *Bryonia* to woodcarvers, architects and designers, especially in the Gothic period.

Bryonia is in flower from May to September. Unlike most plants it does not produce male and female parts in the same flower, but individuals are either male or female. The second part of the Latin name, *dioica*, comes from the term dioecious, meaning 'with two houses'. Male plants bear long-stalked clusters containing three to eight small flowers (12–18mm across) with five creamy-white, yellow-green-veined petals enclosing five stamens (four of which fuse as two pairs, giving the impression there are only three stamens). Female plants produce clusters of two to five very short-stalked flowers that are slightly smaller than the male flowers and bear three Y-forked, downy stigmas.

As in the rest of the melon family the fruit of white bryony develops beneath the petals. It takes the form of a succulent, pea-sized berry. By the time these ripen from green, through yellow and orange to red, the leaves, tendrils and stems have turned brown, leaving the fruits to over-winter like strings of crimson pearls.

There is one part of the plant that is rarely seen, a pity because it is not only why the plant is implicated in some fascinating folklore, but also explains why it is known as white bryony. The white part is the tuberous fleshy root and although the French used to call it 'Navet du Diable'— the Devil's turnip—its knobbly, much-branched shape is more reminiscent of a fat parsnip. Biologically, this is an over-wintering food store that fuels the remarkably fast summer growth of the plant. More interestingly, the root shape of white bryony has resulted in the old local name of mandrake in many parts of Britain.

The true mandrake (*Mandragora officinarum*) is native to Mediterranean areas and the Middle East. Along with other members of the nightshade family mandrake's poisonous alkaloids (especially from its tomato-like fruits) have a history of use for both good and evil. More famous than its poisonous nature is the fact that its large fleshy, branching taproot appears to mimic the human form. One result of this is that the dried root has often been kept as a talisman to promote fertility in barren women. Despite Gerard's rebuttal of this in his 16th-century herbal, the biblical story (Genesis 30:14–16) in which Rachel sought Mandrake to overcome her barrenness, kept the practice alive.

Mandrake roots were highly prized but, not being native to Britain, were very expensive. Fraud was rife and the most common substitute was the swollen root of white bryony. The roots were often carved to create a greater resemblance to the human form. By the time of Henry VIII the roots were drilled with small holes into which millet seeds were placed. These were allowed to germinate before the whole root was dried, resulting in a shape that resembled a hairy humanoid figure. Other tricks involved putting moulds around the growing roots so that they produced the most human-like shapes.

Even as late as the 19th century herbalists' shops would often display dried 'mandrakes' hanging from the ceiling. In 1916 an Oxford labourer presented the curator of the Pitt Rivers Museum with a 16 inch bryony root that he believed to be mandrake. Bryony roots have also been responsible for poisoning cattle, although in former times they were used medicinally. In the 14th

White bryony

White bryony   (a) Male flowers   (b) Fruits   (c) Leaf and female flowers

century white bryony was deemed a cure for leprosy and was later used both by doctors and vets as a purgative and diuretic. The plant was also involved in the treatment of rheumatism and whooping cough. An extract of the fresh root of *Bryonia* is still used in homoeopathic medicine for the treatment of fevers and rheumatism.

White bryony is not distributed over the whole of the British Isles, being absent (except where deliberately planted) from Ireland, Scotland and all but the eastern parts of Wales. It is most common in the hedges of southern England. It is absent from the extreme south-west and is uncommon north of the Midlands, although it does extend up the east coast as far as Northumberland. It is most at home at low altitudes in hedges and scrubland on well-drained, base-rich soils.

Some years ago, Colin Ennis, a good friend of mine, received a call from his former university as part of a fund-raising exercise. The woman on the phone introduced herself as Bryony, to which Colin replied, 'Are you a white bryony or a black bryony?' It was only during the ensuing silence that Colin realised the potential racial overtones of his flippant botanical question. Colin, being a man of many parts, was only showing off his knowledge—for there are indeed two native British plants: white bryony and black bryony.

Black bryony, or *Tamus communis* to give it its scientific name, is frequently confused with white bryony. *Tamus* is also a fast-growing, poisonous, red-berried, non-woody climber growing in the hedgerows of the southern half of Britain. The main distinguishing features are that black bryony is totally hairless, it does not bear tendrils (instead the stem twines in a clockwise direction round surrounding plants) and its alternate, long-stalked, large, broadly heart-shaped leaves are

124

dark green and glossy. The main leaf veins are palmate but the side veins are netted which, together with the broad leaves, are unusual features in a plant that is classified as a mono-cotyledon and is closely related to the lily family.

Whereas white bryony is our sole native member of the melon family, black bryony is the only native British species in the yam family, most members of which are tropical climbing plants with large underground tubers. The tubers of yams and sweet potatoes provide valuable food, but those of *Tamus* are poisonous. The outer skin of the tuber is a dark brownish-black, thus explaining the black part of the plant's name.

Black bryony has another feature in common with its white namesake in that it too is dioecious. Populations of *Tamus* tend to contain more male than female plants. The tiny yellow-green flowers are only about half as big as those in *Bryonia* and contain six petals. The female flowers (with three two-lobed stigmas) occur in small, short-stalked hanging clusters, in contrast to the upright slender spikes bearing 20 or more male flowers, each with six stamens. The poisonous, shiny red berries often persist into the winter, despite the rest of the above-ground parts dying back. Gerard (1597), in explaining the name, mistakenly described the berries as black; he had obviously not observed the plant!

Both bryony species are attractive, if potentially dangerous additions to our hedgerows, where the colour of their vivid red berries defies their black and white image.

Black bryony

Mandrake-like root of white bryony

125

Wild Arum

Arum maculatum

Many years ago I was interviewed for the job as presenter of a natural history television programme aimed at young people. Every interviewee had to prepare a five-minute presentation about a British animal, bird or (in my case) plant. My chosen subject was wild arum. I didn't get the job, partly because I required far more than five minutes to do justice to such a fascinating plant.

Wild arum is one of dozens of local names for *Arum maculatum* and is one of the few polite ones. Most of the other names have a sexual connotation on account of the unusual appearance of the plant when in flower. These include lords and ladies, bulls and cows, jack in the pulpit and the highly descriptive dog's dick. Cuckoo pint (and this should rhyme with mint) is a cleaned up version of the older name cuckold's pintle; the latter term was a former slang word for a penis.

Even the apparently innocuous name wake robin is unlikely to have anything to do with the red-breasted bird, but rather concerns the sexual antics of the house goblin Robin Goodfellow. The plant was formerly considered to be an aphrodisiac, as alluded to in the 17th-century play *Loves Metamorphosis*: 'they have eaten so much Wake Robin that they cannot sleep for love'. In country districts, even as late as the 1920s, girls were loath to touch the plant for fear of becoming pregnant.

The specific Latin name *maculatum* means spotted or with blemishes (the opposite of immaculate) and refers to the purple-black blotches on the leaves. As with other plants with spotted leaves it was believed (especially in Wales) that it had grown at the foot of the cross where its leaves had been splattered with the blood of Christ. Wild arum is a very variable species and many plants have unspotted leaves. In 1640 Parkinson listed two species, and placed the plain-leaved form under the *Arum non-maculatum*. It was later realised that as with different hair colour in people, the spotted and unspotted forms were simply examples of variation within the same species. Prime's classic work, published in 1960, showed that generally the spotted form is the less common one and that populations in the south-east of England have a much higher percentage of spotted individuals than do populations in Ireland and the extreme north of England.

Wild arum is common throughout lowland Britain, but is less frequent in the north and is doubtfully native in northern Scotland. It is a plant of moist soils in shady habitats—woods, copses and hedgerows—where it often forms large patches. The plant over-winters as an underground rhizomatous tuber from which the leaves emerge during February and March. The long-stalked, glossy, dark green, net-veined leaf blades are typically triangular or arrow-shaped in outline. A stalked cowl-like spathe emerges from among the overlapping leaf bases from April onwards. This modified leaf is of a lighter green than the true leaves and is often suffused with

Arum fruits

purple. The margins at the spathe base overlap to enclose a chamber from which a finger-like organ known as the spadix emerges. This is normally a purple-brown colour, but in some plants the spadix is pale yellow. More enigmatically, in some plants the spathe bases overlap left over right while in others the overlap is right over left.

The actual flowers are hidden in a chamber at the base of the spadix. The enclosure entrance is partially blocked by a double ring of bristles, below which is a zone of red-brown stamens. The ovaries, arranged like the seeds of corn on the cob, are below the stamens. Chemical changes in the spadix cause it to heat up and emit a strange smell that attracts tiny flies known as owl midges. These become temporarily trapped within the flowering chamber and later carry pollen to another plant, thus ensuring that arum is cross-pollinated. By the autumn the leaves, spathe and spadix wither away, leaving a lollipop-like cluster (up to 5cm long) of orange-red berries at the stalk apex. Despite their acrid taste, the berries are attractive to children, but their consumption results in violent vomiting and diarrhoea.

The rest of the plant (including the tuber) also contains a range of poisonous chemicals, which are destroyed by heat. The tubers have a very high starch content and this has resulted in their historic use by the laundry trade. The need for laundry starch reached its zenith during Elizabethan times. Stiffened dresses and three-tiered ruffs relied on regular starching, much of the starch originally coming from wheat. During the 16th century there was rarely enough wheat to feed the increasing population of Britain and alternative sources of starch were sought for the laundry industry. Gerard reported in 1597, 'The most pure and white starch is made from the roots of Cuckow-pint; but most hurtfull to the hands of the Laundresses that have the handling of it …'.

Wild arum showing the tubers

Two-hundred years later, when the use of *Arum* had been consigned to history, wheat was again in short supply and The Royal Society of Arts offered a prize of 30 guineas 'for discovering a method of manufacturing starch from material not used as food by man'. The prize was claimed by a woman from the Isle of Portland who revived the former use of the 'roots' of *Arum maculatum*. Even so, starch produced this way was expensive and as Gerard had noted it contained a skin irritant, so it was not popular with laundry workers. By the middle of the 19th century the quarrying of stone at Portland had reduced the land area from which the plant could be collected and the production of starch had all but ceased. Another local use of the tubers also died out by about 1850. This was the production (including baking and washing to remove the poisons) of Portland arrowroot, also known as Portland sago, which was sold as a food for invalids and those recovering from operations.

In my youth I remember my mother using starch on my father's shirts. The brand she used was Robin Starch and it had a picture of the red-breasted bird on the packet. I strongly suspect that the robin name was more likely to have originated from wake robin, one of the many names for *Arum maculatum,* even if by then the plant no longer supplied the starch in the packet. In the past Parisiennes used a facial cosmetic called Cyprus powder made from *Arum* tubers in a manner similar to the production of Oris powder from *Iris*. The plant still provides a homoeopathic remedy for the treatment of gastritis. Aristotle's writings include the fascinating legend that bears used *Arum* to soothe their paws after hibernation, during which time hunger forced them to chew their feet!

In 1854 Mr Hambrough recorded what is now called rare lords and ladies (*Arum italicum*) growing near his home on the Isle of Wight. This species has since been recorded close to the south coasts of Wales, Sussex, Hampshire, Dorset, Devon and Cornwall. It also occurs on the Isles of Scilly, where *Arum maculatum* is absent. On much of the southern side of the island of Portland *Arum italicum* is the more common species. Given that the tuber of *Arum italicum* is four times as big as that of *Arum maculatum*, it appears that the starch and arrowroot industry succeeded in Portland partly owing to the local availability of much larger tubers.

Arum italicum is a larger plant than its common relative, reaching a height of 50cm. It produces up to eight longer stalked leaves from late September and these stay green over winter, when *maculatum* has no leaves. The basal leaf lobes converge and may even overlap, while the leaf veins are paler than the rest of the blade, which is rarely spotted (and if so the spots are very small). Despite being much earlier into leaf *italicum* does not come into flower until May or June; later than *maculatum*. The large yellow spathe (up to 35cm long) frequently droops at its tip and is never purple blotched. The spadix is always yellow. The cluster of orange-red fruits is at least twice as long as in *maculatum*.

Arum plants are frequently grown in gardens. A common garden-grown arum is a different subspecies of *Arum italicum*. The most popular cultivar is 'Marmoratum', which has similar features to our native subspecies and occasionally becomes naturalised, when it can be distinguished from the native form described above by its leaves having divergent lower lobes and creamy-white veins that give a marbled effect. A further subspecies (*albispathum*), also grown in gardens, has smaller plain green leaves and white spathes. Related species in the genus *Zantedeschia*, including the white-spathed arum lily, are becoming increasingly popular as pot plants and cut flowers.

Cuckoo-Pint

A brown matchstick held up in the wind,
the bract-leaf cupped around it like a palm.

March had not extinguished it: there it lurked,
sly as something done behind the sheds,

slithering from its half-unrolled umbrella
as we snipped pussy-willow from the lanes.

To come instead on this man of the woods,
tanned and cowled and clammed inside his collar,

his shirt-front spattered with tobacco stains,
his poker oozy with tuber-froth,

was like learning by accident a secret
intended for later, exiting

and obscene and not to be gone back on,
like the knowledge of atoms or death.

Blake Morrison

An immaculate wild arum

Wormwood

Artemisia absinthium

In 1951 Edlin described this plant as 'A perennial of dusty and untidy aspect'. Today it might be described as having a bad hair day. Wormwood is but one of a large number of plants found in the genus *Artemisia*, of which the only other common British native is mugwort (*Artemisia vulgaris*). Cultivated species include Russian tarragon (*Artemisia dracunculus*), southernwood (*Artemisia abrotanum*) and Roman wormwood (*Artemisia pontica*). Many species of *Artemisia* are highly aromatic and have a fascinating history of medicinal, culinary and other uses. There is some disagreement over the origin of the generic name and some authors have copied Gerard's explanation that it was taken from Artemisia who, as queen to King Mausolus, built the famous Mausoleum (c.750 BC). The much more likely source, given the uses to which members of the genus are put, is Artemis, the Greek goddess of hunting, who looked after women in childbirth. On a recent visit to Jordan my local guide pointed out a species of *Artemisia* and explained that local hunters had rubbed themselves with the aromatic herb to disguise their human scent and thus get closer to their quarry. Game (especially!), set and match to Artemis.

Wormwood is an erect, tufted, very aromatic (not particularly pleasant smelling), perennial growing to about one metre tall. The alternate leaves vary in shape from the lowest, which are long-stalked and finely divided, to those near the apex, which are unstalked and undivided. Unlike those in the similar but much less aromatic mugwort, the leaf segments of wormwood are blunt-ended and the leaves are covered on both sides with silky white hairs. The stem is often woody at its base, but green and angular among the new growth. In mid-summer wormwood bears hundreds of tiny, cup-shaped, drooping, pale yellow flower heads. In contrast, mugwort has alternating vertical stripes of green and purple on its stem and produces brownish flower heads.

Wormwood is probably native, although some authors consider it to be a very early introduction. A plant of disturbed ground, old quarries, roadsides and coastal regions, it is much less frequent north of Yorkshire. *Artemisia* is especially common in urban situations as a weed of carpark margins and demolished factory sites, where it is frequently found growing with Oxford ragwort (see pages 94–97). It is susceptible to frost damage and this may explain why it is more common in southern, coastal and urban regions of Britain.

Just as there is some doubt as to the origin of the Latin name for the genus (we will explore the specific name later), there is also confusion over the English name. The Old English name was *wermod*, much closer to the modern German *wermut* and the French *vermouth*. Such spellings have nothing to do with worms, but the plant has! Species of *Artemisia* were used by the Ancient Egyptians to expel worms and Pliny writes that extracts of wormwood were used against intestinal worms. More modern texts also extol its anthelmintic (worm-dispelling) properties, so it is perhaps not surprising that the English spelling changed from werm to worm. Unfortunately, in some localities it was called mugwort, a confusion that has also crept into the story of the Russian name for wormwood. *Chernobyl* (actually chernobil) is described as 'bot(any) moxa, mugwort, artemisia'.

The bitter legacy left by the more famous Chernobyl is mirrored by wormwood and some of its relatives. The Bible makes several references to wormwood (probably *Artemisia judaica*) and highlights its bitterness. The bitterness of our own species has been the basis for another important medicinal use, namely as an aperitif that stimulates the digestive juices. Bitter substances enhance the secretion of bile, an enzyme produced in the liver and stored in the gallbladder. Bile aids in the digestion of fatty foods in the small intestine. The bitter taste of wormwood was previously put to good use as a flavouring (and preservative) for ale, before the use of hops that are now used in the production of best bitter. Shakespeare, in *The Merry Wives of Windsor*, makes reference to purl, a fortified ale that included *Artemisia* extracts.

The invention of steam distillation resulted in not only spirit drinks of a much higher alcohol content than that found in ordinary wines or beers, but also the ready extraction of previously insoluble plant chemicals. The high alcohol content and plant extracts were neatly combined with the invention of liqueurs. One of these, absinthe, included wormwood (hence the specific name *absinthium*) as an important ingredient and was originally developed in Switzerland at the end of the 18th century. In France a few years later Henri-Louis Pernod was one of a number of people to start production of absinthe. Towards the end of the 19th century this clear green drink had become part of the French way of life, immortalised in early evening 'green hour'. The bitterness of the aperitif and the alcohol level were both partly abated by pouring water on to a sugar cube placed in a sieve-like spoon over a glass containing a small amount of absinthe. The resulting pale yellow opaque liquid was then sipped slowly.

Manet, Degas and Van Gogh all painted pictures on the theme of absinthe; in 1887 Toulouse-Lautrec completed a pastel of Van Gogh with a glass of absinthe and Picasso continued the trend with the *Absinthe Drinker* in 1901. Some years earlier the French poet Baudelaire included the drink in his list of vices. It was an English poet, one Ernest Dawson, who penned the infamous pun 'I understand that absinthe makes the tart grow fonder'. Unfortunately, any possible

Young plant of wormwood

aphrodisiac effect resulting from absinthe was minor when compared with the other symptoms. While artists and others enjoyed the resulting hallucinations, absinthe drinkers also suffered convulsions, paralysis and sleeplessness. As the consumption of absinthe soared in the latter half of the 19th century, scientific articles were published establishing the fact that the neurological effects could not be symptomatic of the alcohol content alone.

It now seems likely that Van Gogh's mental condition, which was to end with his suicide, was exacerbated by his consumption of absinthe. Intriguingly, the drug santonin, extracted from a related species of *Artemisia*, is known to cause xanthopsia (seeing with a yellow hue), which may be one reason for the famous yellow hue of so many of his paintings. Despite the warnings (rebuffed by those whose livelihood depended on absinthe), annual consumption in France topped ten million gallons in the years before World War I. By this time the production of absinthe had been banned in Switzerland, Italy and Belgium, and the French finally followed suit during the war. The drink is still available in Spain and on the black market, but for those unable to exist without the original absinthe, similar-tasting drinks without wormwood (but with extra anise) are produced under brand names such as Pernod and Ricard. Vermouth is still made with a small quantity of Roman wormwood flower heads.

Wormwood contains a range of chemicals including the bitter compound absinthin and an essential oil called thujone. The thujone molecule has structural similarities to that of tetrahydrocannabinol and is the likely cause of absinthe's neurotoxic effects. Thujone is also the chemical responsible for stunning intestinal worms, which are then voided in the faeces. Several other plants, including sage and the white cedar (*Thuja occidentalis*), have been found to contain

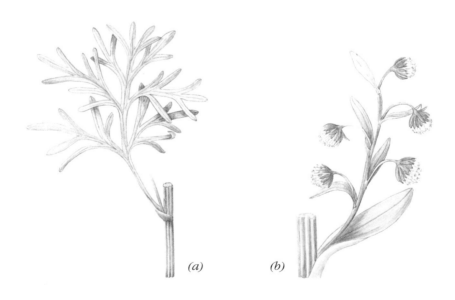

Wormwood   (a) Leaf   (b) Flower heads

thujone. Images of the *Thuja* tree have been made famous by Van Gogh's paintings and this may account for the actions of Van Gogh's doctor, Paul Gachet, who planted one on the artist's grave.

Artemisia absinthium has been put to other uses in addition to the production of aperitifs or worm destroyers. For hundreds of years it was used to promote menstruation and to speed up labour and the expulsion of the afterbirth; uses that the goddess Artemis would no doubt have approved of. It is used by homoeopaths for the treatment of epilepsy and vertigo, while organic gardeners have grown it as a companion plant to ward off insects. Insects were also put off by mugwort, our other native species. Although this, like wormwood, was used to flavour ale, the mug part of the name is more likely to be a derivation of the Anglo-Saxon *myg*, meaning a midge. In Chinese medicine mugwort is used in the cones of moxa that are burnt for heat treatment and cauterisation.

For over 1,000 years the Chinese have used another wormwood relative, *Artemisia annua*, in the treatment of dysentery, scabies and malaria. The active principle artemisinin (the Chinese know it as quinghaosu) has now been isolated and is being taken up by Western medicine to treat those strains of malaria that have become resistant to quinine-based drugs such as chloroquine. A much cheaper synthetic counterpart to artemisinin has recently been developed and is currently undergoing clinical trials.

While Wormwood hath seed, get a handful or twaine,
To save against March, to make flea to refraine:
Where chamber is sweeped, and Wormwood is strowne,
What saver is better (if physick be true)
For places infected than Wormwood and Rue?
It is a comfort for the hart and the braine,
And therefore to have it it is not in vaine.

Thomas Tusser (16th-century farmer)

Hop—another plant
that contains bitters

Yew

Taxus baccata

Three of Britain's native tree species (Scots pine, juniper and yew) belong to a group of plants known as the gymnosperms. These bear naked seeds, unlike those of the angiosperms (flowering plants) in which the seeds are protected by the ovary. Most gymnosperms have small evergreen leaves and bear their seeds in a woody cone; hence the popular term 'conifer'. Yew has small evergreen leaves, but its fruits are more like berries than cones. The Latin *baccatus* means berry-like; *Taxus* is from the Greek word *taxon*, meaning a bow; yew wood is still used in the manufacture of longbows.

Taxus is a shrub or small tree, growing to about 20 metres. The trunk may be multi-stemmed, and is often vertically ribbed and invariably hollow in older trees. The red-brown bark peels away in jigsaw-like pieces and, along with the foliage and seeds, is highly poisonous. Unlike most conifers there is no resin in the bark, leaves or fruits. The young green twigs carry short-stalked, flattened, needle-like leaves, each with a rounded apex that ends in a short point. The 'needles' are dark and shiny green above, displaying a pale stripe either side of the prominent midrib below. The twisted leaf-stalks align the spirally borne needles in a flat, two-ranked configuration on all but the youngest twigs.

Yew is dioecious; individuals are either male or female. I can still picture the look of disbelief coming from a flock of sheep that apparently overheard my comment on a field meeting: 'This is a male yew'. The tree was previously known as ewe and even just 'u'. Male plants produce small globose, soft, pale yellow, catkin-like 'cones' and shed copious amounts of pollen from February to April. On female trees tiny bud-like flowers develop into fruits (arils) resembling small green acorns. By autumn each has ripened to a bright pink-red fleshy 'cup' enclosing a single dark brown seed. A former common name for yew was snotty gogs, an apt description of its strange fruit.

One of my students recalled how, as a small girl, she was repeatedly warned by her mother not to eat the poisonous 'berries' from the yew tree overhanging her garden wall. In an act of rebellion she feasted on the sweet-tasting fleshy fruits, carefully spitting out the hard seeds. She lived to tell the tale because, despite numerous media reports to the contrary (including a recent debate on the letters page of the *Daily Express*), the fleshy aril is the only part of the yew that is not poisonous. As long ago as 1597 Gerard wrote that he had eaten the berries 'without any hurt at all'—some myths are very persistent.

Yew inhabits downland and cliffs as well as woodland, typically on well-drained chalk or limestone soils. It is locally common in England and Wales but rare in Scotland. Yew is also found in churchyards, gardens and parks where, in addition to isolated trees, it is frequently grown as a hedge plant. Nurseries stock yew in a range of cultivars (varieties), including 'Fastigiata', the Irish or Florence Court yew, now more widely planted than the wild form.

The columnar 'Fastigiata' has erect branches bearing spirally arranged leaves. In the 1760s a female plant was discovered by George Willis, a farmer from County Fermanagh. He grew a cutting in his garden and presented another to the nearby Florence Court estate. Seedlings of Irish yew rarely exhibit the unusual growth form, so it is cultivated from rooted cuttings. For this reason all 'Fastigiata' plants, like the original one, are female and bear 'berries'. The columnar yews are especially good subjects for topiary, a garden craze that reached its heights in the 18th and 19th centuries.

The foliage and seeds of yew have a history of poisoning people and animals. Slips of yew were part of the witches' brew in Shakespeare's *Macbeth*. An early 20th-century government publication bluntly reported 'the commonest symptom of yew poisoning is sudden death'. As long ago as 1856 the poisonous alkaloid taxine was extracted from yew. Since then at least 40 similar compounds have been found in a range of yew species, including the Pacific yew (*Taxus brevifolia*), native to North America. The American National Cancer Institute investigated taxol, a chemical synthesised from paclitaxel found in the bark of the Pacific yew. They found it to be one of the most promising of more than 120,000 plant compounds they had tested for anti-cancer properties.

During the 1980s and early 1990s the manufacture of taxol resulted in the destruction of thousands of Pacific yew trees and it was a great relief to conservationists when it was discovered that the European yew (*Taxus baccata*) contained docetaxel, a precursor of paclitaxel, in its foliage. From the 1990s gardeners were paid by pharmaceutical firms to have their yew clippings taken away! In the past ten years similar compounds have been found in the fruits of hazel and in some fungi. Complete synthesis of taxol is possible but too expensive. Taxol is now proving

Yew hedge at Powys Castle

Fleshy arils (berries) of yew

efficacious in the treatment of both breast and ovarian cancers.

As mentioned previously, yew trees were valued for their wood, which made the best longbows. Before the Enclosure Acts, the churchyard was often the only part of a village that was completely fenced off. During my childhood history lessons I learnt that churchyards were the safest places to grow yew wood for the needs of our archers. Naively I wondered why the Church was involved in supplying weapons of war, but that was before I learnt about the Crusades! Research has since shown that most of the yew timber used to make bows in Britain was imported from southern Europe, so why are there so many yews in British churchyards?

Victorian authors emphasised the links between Christianity and the yew. Along with holly, the evergreen yew symbolised eternal life. Continued growth of ancient trees after the majority of the trunk had rotted was likened to life after death. The yew's white sapwood and red heartwood represented the sacramental body and blood of Christ. The red 'berries' were believed to offer protection against evil and witchcraft. Sprays of the male tree (bearing yellow 'flowers') were sometimes collected from churchyards on Palm Sunday in place of the more frequently used willow catkins. The dark coloured yew was deemed appropriate to a burial site, where the minister would often meet the incoming pall-bearers beneath a yew tree. Yew foliage was even thrown into the grave to mark the soul's eternal nature. Folklore included a belief that yew trees thrived on corpses. Given that bones are rich in calcium and the tree's preferred native habitat is on chalk or limestone soils, the macabre suggestion may not be so wide of the mark.

The vast majority of ancient yews in Britain are found in churchyards, not in the wild. Rather than talk about yew trees planted beside churches it has become fashionable to speculate whether early Christian churches were built beside yew trees, some of which may still be living. In the 1940s Vaughan Cornish argued that yew was sacred to the Druids (as a symbol of immortality) in pre-Christian Britain. He inferred that Christian churches were built on former religious sites, but stated, 'Although tradition ascribes two thousand years and more to some of our venerable yews, the considered opinion of botanical experts is definitely to the contrary … there is no proof that any now standing date back to the time of the Druids, and it is quite unlikely that they do'.

Over 50 years later, scientific opinion has swung towards the belief that in some sites the yew could well pre-date even the earliest Christian church. Problems about calculating the lifespan of yews remain, not least because the oldest trees have hollow trunks, making ring-count ageing impossible. Equally, growth rates of older yew trees (over 300 years) are too variable for any accurate assessment of age to be inferred from their girth measurement. However, examination of and extrapolation from the few old yew trees of a known age (from historical records) indicate that trees with very large girths, more than 8 metres, would need at least 1,800 years to reach such a size.

Such large-girthed yews exist, usually close to churches. Records from the building of Fountains Abbey in the 12th century show that the monks used two old hollow yews as shelters. One of these trees was only recently lost. Among the more famous churchyard yews is one at Fortingall in Scotland, which many experts consider to be between 4,000 and 5,000 years old. The myth that Pontius Pilate was suckled beneath the tree (his father having served as a legionary in the area) may even have a basis in truth.

Following a review of the association between yews and churches, Mitchell (1996) concluded: 'The church was built to be near the yew. The site was already sacred, the meeting place of the elders and tribes. The church was placed so as to have the maximum shelter from the indestructible evergreen foliage, usually on the windward side of the porch or approach path.' The arguments about which came first, church or yew, will no doubt continue until a more accurate method for ageing the trees becomes available. Yew wood is renowned for making the best wands: if only Harry Potter could ask a venerable old tree 'how old are you?'

Y was a Yew

Y was a yew,
Which flourished and grew,
By a quiet abode
Near the side of a road.

y!
Dark little Yew!

Edward Lear

Ancient
churchyard
yew

The eternal yew

Bibliography

Aaron, J.R. and Richards, E.G. (1990) *British Woodland Produce*. Stobart Davies.

Acton, J. and Sandler, N. (2001) *Mushroom*. Kyle Cathie.

Allegro, J. (1970) *The Sacred Mushroom and the Cross*. Doubleday.

Baker, M. (1996) *Discovering the Folklore of Plants*. Shire Publications.

Baren, M. (2000) *How it all Began in the Pantry*. Michael O'Mara Books.

Barnard, J. and M. (1988) *The Healing Herbs of Edward Bach*. Bach Educational Programme.

Bean, W.J. (1925) *Trees and Shrubs Hardy in the British Isles*. 4th Ed. John Murray.

Benjamin, D. (1995) *Mushrooms: Poisons and Panaceas*. W.H. Freeman and Company.

Betjamin, J. (1954) In the Licorice Fields at Pontefract from *A Few Late Chrysanthemums*. John Murray.

Bishop, W.J. (1959) *A History of Surgical Dressings*. Robinsons and Sons.

Bisset, N.G. and Wichtl, M. (Eds) (2001) *Herbal Drugs and Phytopharmaceuticals*. 2nd Ed. Medpharm Scientific Publishers.

Blamey, M., Fitter, R. and Fitter, A. (2003) *Wild Flowers of Britain and Ireland*. A & C Black.

Bolton, G (2000) A Christmas Gift from Siberia. *The British Journal of General Practice* December.

Bon, M. (1987) *The Mushrooms and Toadstools of Britain and North-western Europe*. Hodder & Stoughton.

Brickell, C. (Ed.) (1996) *The Royal Horticultural Society A–Z Encyclopedia of Garden Plants*. Dorling Kindersley.

Briggs, R.W. (1993) *'Chinese' Wilson*. HMSO.

Brightman, F.H. (1966) *The Oxford Book of Flowerless Plants*. Oxford University Press.

British Herbal Medicine Association (1976, 1979) *British Herbal Pharmacopoeia Parts 1 & 2*. British Herbal Medicine Association.

British Medical Association (1988) *Guide to Medicines and Drugs*. Dorling Kindersley.

Brough, J.C.S. (Ed.) (1947) *Timbers for Woodwork*. Evans Brothers.

Bunney, S. (Ed.) (1984) *The Illustrated Book of Herbs*. Octopus Books.

Carpenter, H. (2003) *Spike Milligan The Biography*. Hodder & Stoughton.

Chancellor, P.M. (1971) *Handbook of the Bach Flower Remedies*. C.W. Daniel.

Clapham, A.R., Tutin, T.G. and Warburg, E.F. (1962) *Flora of the British Isles*. 2nd Ed. Cambridge University Press.

Cooke, M.C. (1871) *A Plain and Easy Account of the British Fungi*. Robert Hardwicke.

Cooke, R.C. (1977) *Fungi, Man and his Environment*. Longman.

Cornish, V. (1946) *The Churchyard Yew and Immortality*. Frederick Muller.

Courtecuisse, R. and Duhem, B. (1995) *Mushrooms and Toadstools of Britain and Europe*. HarperCollins.

Cox, E.H.M. (1945) *Plant-Hunting in China*. Collins.

Culpepper, N (1653) *The English Physician or The Complete Herbal*. Foulsham.

Drower, G. (2001) *Gardeners, Gurus and Grubs*. Sutton.

Edlin, H.L. (1951) *British Plants and Their Uses*. Batsford.

Evelyn, J. (1664) *Sylva*. Arthur Doubleday.

Fainlight, R. (1997) Autumn Crocus from *Sugar-Paper Blue*. Bloodaxe Books.

Findlay, W.P.K. (1967) *Wayside and Woodland Fungi*. Frederick Warne & Co.

Fischer, D.W. and Bessette, A.E. (1992) *Edible Wild Mushrooms of North America*. University of Texas Press.

Freethy, R. (1985) *From Agar to Zenry*. Crowood Press.

Freethy, R. (1987) *British Ferns*. Crowood Press.

Gerard, J. (1597) *The Herball, or Generall Historie of Plants*. London.

Gilbert, O. (2000) *Lichens.* HarperCollins.

Graves, R. (1960) *The Greek Myths.* Cassell & Co.

Green, T. (2005) Is there a case for the Celtic Maple or the Scots Plane? *British Wildlife* 16.3.

Gregory, P. (1998) *Earthly Joys.* HarperCollins.

Gregory, P. (1999) *Virgin Earth.* HarperCollins.

Grieve, M. (1931) *A Modern Herbal.* Jonathan Cape.

Griggs, B. (1981) *Green Pharmacy.* Jill Norman & Hobhouse.

Grigson, G. (1955) *The English Man's Flora.* Phoenix House.

Grigson, G. (1974) *A Dictionary of English Plant Names.* Allen Lane.

Grime, J.P., Hodgson, J.G. and Hunt, R. (1990) *The Abridged Comparative Plant Ecology.* Chapman & Hall.

Harding, P. (1996) *Collins Gem Mushrooms.* HarperCollins.

Harding, P. (2004) *The Magic of Christmas.* Metro for John Blake.

Harding, P., Lyon, A. and Tomblin, G. (1996) *How to Identify Edible Mushrooms.* HarperCollins.

Harding, P. and Oxley, V. (2000) *Wild Flowers of the Peak District.* Hallamshire Press.

Harding, P. and Tomblin, G. (1998) *How to Identify Trees* HarperCollins.

Hartzell, H., Jr (1991) *The Yew Tree A Thousand Whispers.*

Hawksworth, D.L., Kirk, P.M., Sutton, B.C. and Pegler, D.N. (1995) *Ainsworth & Bisby's Dictionary of the Fungi.* CAB International.

Hoffmann, D. (1983) *The Holistic Herbal.* Findhorn Press.

Hollman, A. (1991) *Plants in Medicine.* The Chelsea Physic Garden.

Houseman, P.A. (1944) *Licorice. Putting a Weed to Work.* The Royal Institute of Chemistry of Great Britain and Ireland.

Hudler, G.W. (1998) *Magical Mushrooms, Mischievous Molds.* Princeton University Press.

Hughes, T. (1967) Fern from *Wodwo.* Faber and Faber.

Hughes, T. (1986) Sunstruck Foxglove from *Flowers and Insects.* Faber and Faber.

Hulme, F.E. (undated) *Familiar Wild Flowers.* Cassell.

Hyde, H.A. and Wade, A.E. (1940) *Welsh Ferns.* National Museum of Wales.

Johns, C.A. (1903) *The Forest Trees of Britain.* 9th Ed. S.P.C.K.

Johnson, O. and More, D. (2004) *Collins Tree Guide.* HarperCollins.

Jordan, M. (1976) *A Guide to Wild Plants.* Millington.

Kennedy, C.E.J. and Southwood, T.R.E. (1984) The number of species of insects associated with British trees. A reanalysis. *Journal of Animal Ecology*: 53.

Keys, J.D. (1976) *Chinese Herbs* Charles E. Tuttle.

King, A. and Clifford, S. (1989) *Trees Be Company.* Bristol Press for Common Ground.

Kreig, M.B. (1965) *Green Medicine—The Search for Plants that Heal.* George G. Harrap.

Lear, E. (1947) Y was a Yew from *The Complete Nonsense of Edward Lear.* Faber and Faber

Lehane, B. (1977) *The Power of Plants.* McGraw-Hill.

Leith-Ross, P. (1984) *The John Tradescants.* Peter Owen.

Lewington, A. (1990) *Plants for People.* Natural History Museum Publications.

Llewellyn-Williams, H. (1987) Elder (Ruis) from *The Tree Calendar.* Poetry Wales Press.

Longman (1968) *Longman's English Larousse.* Longman.

Mabey, R. (1977) *Plants with a Purpose.* Collins.

Mabey, R. (Ed.) (1988) *The Complete New Herbal.* Elm Tree Books.

Mabey, R. (1996) *Flora Britannica.* Sinclair-Stevenson.

McClintock, D. (1966) *Companion to Wild Flowers.* G. Bell and Sons.

McClintock, D. and Fitter, R.S.R. (1956) *The Pocket Guide to Wild Flowers.* Collins.

Manniche, L. (1989) *An ancient Egyptian Herbal.* British Museum Publications.

Matossian, M.K. (1989) *Poisons of the Past—Molds, Epidemics and History.* Yale University Press.

Merryweather, J. and Hill, M. (1992) The fern guide. *Field Studies*, 8, 101–188.

Miles, A. (1999) *Silva—The Tree in Britain*. Ebury Press.

Mitchell, A. (1974) *A Field Guide to the Trees of Britain and Northern Europe*. Collins.

Mitchell, A. (1996) *Alan Mitchell's Trees of Britain*. HarperCollins.

Morrison, B. (1999) Cuckoo-Pint from *Selected Poems*. Granta Books.

Nicholson, N. (1994) Weeds from *Collected Poems*. Faber and Faber.

Ody, P. (1993) *The Herb Society's Complete Medicinal Herbal*. Dorling Kindersley.

Over, L. (1987) *The Kelp Industry in Scilly*. Isles of Scilly Museum Publication.

Parkinson, J. (1629) *Paradisi in Sole Paradisus Terrestris*. Humphrey Lownes and Robert Young.

Parkinson, J. (1640) *Theatricum Botanicum*. London.

Paulin, T. (1983) The Book of Juniper from *Liberty Tree*. Faber and Faber

Pavord, A. (1999) *The Tulip*. Bloomsbury.

Pegler, D. (1990) *Field Guide to the Mushrooms and Toadstools of Britain and Europe*. Kingfisher.

Phillips, R. (1981) *Mushrooms and Other Fungi of Great Britain and Europe*. Pan.

Plath, S. (1960) Mushroom from *The Colossus and Other Poems*. Harper and Row.

Plath, S. (1971) Winter Trees from *Winter Trees*. Faber and Faber.

Press, J.R., Sutton, D.A. and Tebbs, B.M. (1981) *Field Guide to the Wild Flowers of Britain*. Reader's Digest.

Prime, C.T. (1960) *Lords and Ladies*. Collins.

Proctor, M., Leo, P. and Lack, A. (1996) *The Natural History of Pollination*. Paperback ed. HarperCollins.

Reader's Digest (1981) *Field Guide to the Wild Flowers of Britain*. Reader's Digest.

Rose, F. (1981) *The Wild Flower Key*. Frederick Warne.

Rudgley, R. (1993) *The Alchemy of Culture—Intoxicants in Society*. British Museum Press.

Schauenberg, P. and Paris, F. (1977) *Guide to Medicinal Plants*. Lutterworth Press.

Shephard, S. (2003) *Seeds of Fortune—A Gardening Dynasty*. Bloomsbury.

Silkin, J. (1965) Moss from *Nature with Man*. Chatto and Windus.

Sitwell, S. (1982) Tulip Tree from *An Indian Summer: 100 Recent Poems*.

Stace, C. (1991) *New Flora of the British Isles*. Cambridge University Press.

Stearn, W.T. (1992) *Stearn's Dictionary of Plant Names for Gardeners*. Cassell.

Stearn, W.T. (1992) *Botanical Latin*. 4th Ed. David & Charles.

Swain, T. (Ed.) (1972) *Plants in the Development of Modern Medicine*. Harvard University Press.

Tennyson, A. (1869) Flower in the Crannied Wall from *The Holy Grail and Other Poems*.

Thistleton Dyer, T.F. (1889) *The Folk-Lore of Plants*. Chatto & Windus.

Thomson, W.A.R. (1976) *Herbs that Heal* Adam and Charles Black.

Vickery, R. (1995) *Oxford Dictionary of Plant Lore*. Oxford University Press.

Wasson, R.G. (1971) *Soma: Divine Mushroom of Immortality*. Harcourt Brace Jovanovich.

Watson, E.V. (1968) *British Mosses and Liverworts*. Cambridge University Press.

Wheelwright, E.G. (1935) *The Physick Garden: Medicinal Plants and their History*. Houghton Mifflin.

Wilks, J.H. (1972) *Trees of the British Isles in History and Legend*. Frederick Muller.

Wilson, E.H. (1913) *A Naturalist in Western China*. Methuen.

Wordsworth, W. (1807) To the Small Celandine from *Poems in Two Volumes*.

Index

Abortifacient   63
Absinthe   131
Acer Pseudoplatanus   111–113
Acetone   57
Aesculus hippocastanum   54–57
Algae   14
Alkaloids   27, 40, 83, 119, 135
Allegro, John   46
Alnwick Castle   26
Amanita caesarea   47
Amanita muscaria   46–49
American Elder   34
Amistar   104
Antabuse   108
Aphids   110
Arctium   50
Aristotle   128
Armillaria   102
Artemisia absinthium   130–133
Artemisia annua   133
Artemisia vulgaris   130
Arum italicum   128–129
Arum maculatum   126–129
Asthma   22, 70, 89
Athyrium filix-femina   79
Atropa belladona   26–29
Atropine   27, 28
Auricularia auricula-judae   36
Autumn Crocus   82

Bach, Edward   57
Badedas   57
Banks, Joseph   59
Bassett's   71
Beard lichens   92
Beer   72
Betjamin, John   73
Bile   22, 131
Bitters   22, 63, 72, 119, 132
Black bryony   124
Bladder Wrack   14–18
Blood clotting   89

Bog Moss   18–21
Bog Rosemary   19
British Mycological Society   102
Bryonia dioica   122–125
Bryophyte   18
Buckler Fern   79
Burdock   50
Bushmills whiskey   20
Butterbur   24

Caesar's Mushroom   47
Cancer   135
Carbenoxolone   72
Cardiac glycosides   52
Castanea sativa   55
Centranthus ruber   118–119
Chamaemelium nobile   100
Chamomile   100
Chaucer   70
Chelidonium majus   66, 68
Chernobyl   131
Chesterfield   119
China   30
Chipping Camden   31
Chopping boards   112
Christianity   46, 136
Cigarettes   22, 72, 89
Cinnebar moth   96
Claviceps purpurea   38–41
Clusius, Carolus   54
Cochicum autumnale   82–85
Colchicine   83
Coleraine   94
Coltsfoot   22–25
Comfrey   50
Common ink cap   107
Conkers   56
Cooke, Mordecai   48, 107
Cooke, Rod   74
Coprinus atramentarius   107
Coprinus comatus   106–109
Coprinus picaceus   108

Coughwort   22
Coumarins   88, 101
Crippen   28
Cuckoo pint   126
Culloden   94
Culpeper, Nicholas   44, 66
Cymbalaria muralis   58–61
Cymbalaria pallida   60
Cystisis   63

David, Jean Pierre   30
Davidia involucrata   30
Deadly Nightshade   26–29
Delavay, Jean Marie   31
Dendrochronology   40
Derbyshire neck   17
Digitalis purpurea   50–53
Digoxin   52
Dioscorides   44, 66, 80, 120
Diuretic   36, 52, 63
Doctrine of Signatures   24, 43, 66, 68, 81
Dodgson, Charles   48
Dove Tree   30
Dropsy   52
Druids   136
Dryopteris dilatata   79
Dryopteris filix-mas   78
Dwarf Elder   34
Dyes   36

Egypt   63, 70
Elder   34
Eleusis   41
Emerson, Ralph Waldo   101
Ergometrine   40
Ergot   38–41
Ergotamine   41
Essential oils   45, 62, 89, 101, 119, 132
Euphrasia confusa   42
Euphrasia nemorosa   42

Euphrasia officinalis   42–45
Evelyn, John   110, 112, 114
Evernia prunastri   90–93
Evolution   97
Eyebright   42–45

Fainlight, Ruth   85
Fairies   50, 52
Father Christmas   48
Feldene   83
Fig   66
Fig-mulberry   112
Figwort   66
Fire extinguisher   72
Flammulina velutipes   103
Florence Court yew   134–135
Fly Agaric   46–49, 74, 76, 106
Formication   39
Fortingall   137
Fortune, Robert   30
Fountains Abbey   26, 137
Foxglove   50–53
Frog   66
Frond   78
Fucus serratus   15
Fucus spiralis   15
Fucus vesiculosus   14–18
Furness Abbey   26

Genesis   123
Gerard, John   23, 24, 26, 54, 82,
   86, 88, 118, 123, 125, 130, 134
Gin   63
Glass making   16, 81
Glycyrrhiza glabra   70–73
Gout   83
Grasmere   69
Graves, Robert   48
Greater Celandine   66, 68
Gregory, Philippa   54
Grigson   34, 62
Groundsel   96
Gunpowder   63
Gymnosperm   62, 134

Hallucinations   48, 74, 76
Hampton Court   54
Handkerchief Tree   30
Harry Potter   27
Henry, Augustine   31
Homoeopathy   124, 133

Horse Chestnut   54–57
Hughes, Ted   53, 81
Huxley, Aldous and Anthony   46

Ice Age   62, 64
Iodine   17
Ion-exchange resin   19
Irritable bowel syndrome   121
Isles of Scilly   15, 22, 51, 60, 128
Ivy-leaved Toadflax   58–61, 94

Jesus   63, 126
Jew's Ear Fungus   36
Juniper   62–65
Juniperus chinensis   64
Juniperus communis   62–65
Juniperus sabina   64

Kelps   15
Kerr, William   64

Lady Fern   79
Lawyer's wig   106
Lear, Edward   137
Leary, Timothy   74
Lesser Celandine   66–69
Liberty cap   74
Lichen   90
Linaria vulgaris   58
Linnaeus, Carl   102
Liquorice   70–73
Liriodendron chinense   117
Liriodendron tulipifera
114–117
Little Women   29
Llewellyn-Williams, Hilary   37
Longbows   134
Lords and ladies   126
LSD   38, 41

Mabey, Richard   82
Madagascar periwinkle   84
Magic Mushroom   74–77, 106
Magnolia   114
Magpie fungus   108
Malaria   133
Male Fern   78
Mandragora officinarum   123
Mandrake   123
Maple, field   112
Matricaria discoidea   98–101

Matricaria recutita   101
Meadow Saffron   82–85
Melilot   86–89
Melilotus alba   86
Melilotus altissima   86
Melilotus officinalis   86–89
Milligan, Spike   72
Milton, John   45
Morgan Motors   116
Morrison, Blake   129
Moss   21, 90
Moss, Kate   93
Mucilage   22
Mugwort   130
Mullein   50
Myxomatosis   62

Nicholson, Norman   25

Oak Moss   90–93
Oudemansiella mucida   102–
   105
Oudemansiella radicata   103
Oxford Ragwort   94–97
Oxford   59, 94, 111

Parkinson, John   23, 54, 58, 85,
   111, 126
Parkinson's Disease   29
Parmelia saxatilis   90
Paulin, Tom   65
Peat   20
Perfumes   93
Pernod   131
Pied Piper   118
Piles   66, 89
Pineapple weed   98–101
Pith   35
Plane   112
Plantlife   62
Plath, Sylvia   109
Pliny   22, 23, 66, 70
Poached egg fungus   102
Polyploidy   84
Pontefract   70
Pontius Pilate   137
Poplar   114
Porcelain fungus   102–105, 106
Portland sago   128
Powell, Baden   49
Pseudoevernia furfuracea   93

143

Psilocybe mexicana 74
Psilocybe semilanceata 74–77
Psilocybin 74, 76

Queen Mother 54

Ragwort 95
Ranunculus ficaria 66–69
Red-berried Elder 34
Resins 63
Revers, Dr 72
Robin starch 128
Romans 63, 70
Rooting shank 103

Saffron 82
Sagina 18
Salem 40
Salicornia 16
Salmon 44
Salt tolerance 99
Sambucus canadensis 34
Sambucus ebulus 34
Sambucus nigra 34
Sambucus racemosa 34
Sceted Mayweed 101
Scrophularia nodosa 66
Scurvy 60
Senecio jacobaea 95
Senecio squalidus 94–97
Senecio vulgaris 96
Shaggy ink cap 106–109
Shakespeare 17, 28, 35, 36, 111,
   113, 131, 135
Sheffield 70, 74
Shroom shops 76
Silkin, Jon 21
Sitwell, Sacheverell 117
Soda ash 15
Solanaceae 26
Solanum dulcamara 26

Sphagnum 18–21
Spiral Wrack 15
Squirrel 112
St Anthony's Fire 39
St John 81
Steroids 72
Strobilurins 104
Sulphur dioxide 91
Surgical dressings 17
Swarfega 22
Sweet Chestnut 55
Sycamore 110–113
Sycamore 110–113, 92
Sylvia Plath 33
Symbiosis 90
Symphytum 50

Tamus communis 124
Tannins 45, 66, 72, 89, 101, 118
Tarragon 130
Taxol 135
Taxus baccata 134–138
Taxus brevifolia 135
Tennyson, Alfred 61
Theophrastus 44, 70, 80
Thoreau, Henry 104
Tinder 24
Tolpuddle 112
Toothed Wrack 15
Tradescant, John 54, 114
Tree Calendar 37
Tuberculosis 52, 66
Tulip tree 114–117
Tulip 114
Turner, William 50, 66, 82, 86
Tusser, Thomas 133
Tussilago farfara 22–25

Ulcers 72
Underdog 110
Usnic acid 93

Valerian 118–121
Valeriana officinalis 118–121
Valium 121
Van Gogh, Vincent 131–132
Veitch, Harry 31
Velvet shank 103
Verbascum thapsus 50
Vermifuge 80, 130
Violin 112
Virginia 114

Wall Valerian 118–119
Ward, Nathaniel 78
Warfarin 89
Wasson, Gordon and Tina 46,
   74
Weeds 25
Weizmann, Professor 57
Well dressing 90
White Bryony 122–125
Wild Arum 126–129
Wild Cranberry 19
Wilson, Ernest 31, 116
Winter trees 33
Witches 27, 28, 34, 40, 52, 63,
   113, 135
Withering, William 52
Woody Nightshade 26
Wordsworth, William 69
World War 20, 22, 23, 28, 57, 85,
   94, 132
Wormwood 130–133
Wound dressings 20
Wracks 14–18

Yew 134–138
York groundsel 97